An Introduction to Plant Breeding

d if the item is requested

An Introduction to Plant Breeding

Jack Brown
Peter D.S. Caligari

Blackwell
Publishing

Blackwell Publishing editorial offices:
Blackwell Publishing Ltd, 9600 Garsington Road, Oxford OX4 2DQ, UK
 Tel: +44 (0)1865 776868
Blackwell Publishing Professional, 2121 State Avenue, Ames, Iowa 50014-8300, USA
 Tel: +1 515 292 0140
Blackwell Publishing Asia Pty Ltd, 550 Swanston Street, Carlton, Victoria 3053, Australia
 Tel: +61 (0)3 8359 1011

First published 2008 by Blackwell Publishing Ltd

2 2009

ISBN: 978-1-4051-3344-9

Library of Congress Cataloging-in-Publication Data
Brown, Jack Houghton.
 An introduction to plant breeding / Jack Brown and Peter D.S. Caligari.
   p. cm.
 Includes bibliographical references and index.
 ISBN: 978-1-4051-3344-9 (pbk. : alk. paper)
 1. Plant breeding. I. Caligari, P. D. S. (Peter D. S.) II. Title.
 SB123.B699 2006
 631.5′2–dc22
                     2006019915

A catalogue record for this title is available from the British Library.

Set in 10/12pt Agaramond
by Newgen Imaging Systems (P) Ltd., Chennai, India
Printed and bound in Great Britain by
TJ International Ltd, Padstow, Cornwall

The publisher's policy is to use permanent paper from mills that operate a sustainable forestry policy,
and which has been manufactured from pulp processed using acid-free and elementary chlorine-free
practices. Furthermore, the publisher ensures that the text paper and cover board used have met
acceptable environmental accreditation standards.

For further information on Blackwell Publishing, visit our website:
www.blackwellpublishing.com

1006017167

Contents

1

Introduction

REQUIREMENTS OF PLANT BREEDERS

The aim of plant breeding is to develop superior cultivars, which are adapted to specific environmental conditions and suitable for economic production in a commercial cropping system.

The basic concept of varietal development is very simple and involves three distinct operations:

- produce or identify genetically variable germplasm;
- carry out selection procedures on genotypes from within this germplasm to identify superior genotypes with specified characteristics;
- stabilize and multiply these superior genotypes and release cultivars for commercial production.

The general philosophy underlying any breeding scheme is to maximize the probability of creating, and identifying, superior genotypes which will make successful new cultivars. In other words they will contain all the desirable characteristics/traits necessary for use in a production system.

Plant breeders can be categorized into two types. One group of plant breeders is employed within private companies, while the other group works in the public sector (e.g. government funded research institutes or universities). Private sector and public sector breeders have different approaches to the breeding process. Many of the differences that exist between public and private breeding programmes are related to the time available for variety release, types of cultivar developed and priorities for characters in the selection process. For example, breeders within the public sector are likely to have a number of responsibilities, related to academic activities or extension services, as well as producing new varieties. Private sector plant breeders tend to have a more clearly defined goal, developing new cultivars and doing it as quickly as possible. In addition, many private breeding organizations are, or are associated with, agrochemical companies. As a result varietal development may be designed to produce cultivars suitable for integration with a specific production system.

Despite the apparent simple description of the breeding process given above, in reality plant breeding involves a multidisciplinary approach. Irrespective of whether a breeding scheme is publicly or privately managed, a successful plant breeder will require knowledge in many (if not all) of the following subjects:

Evolution It is necessary to have knowledge of past progress in adapting crop species if additional advances are to continue into the future. When dealing with a crop species, a plant breeder benefits from knowledge of the time scale of events that have modelled the given crop. For example, the time of domestication, geographic area of origin and prior improvements are all important and will help in setting feasible future objectives.

Botany The raw material of any breeding scheme is the available germplasm (lines, genotypes, accessions, etc.) from which variation can be generated. The biological relationship, which exists within a species and with other species, will be a determining factor indicating germplasm variability and availability.

Biology Knowledge of plant biology is essential to create genetic variation and formulate a suitable breeding and selection scheme. Of particular interest are modes of reproduction, types of cultivar and breeding systems.

Genetics The creation of new cultivars requires manipulation of genotypes. The understanding of genetic procedures is therefore essential for success in plant breeding. Genetics is an ever developing subject but knowledge and understanding that is particularly useful will include single gene inheritance, population genetics, the likely frequencies of genotypes under selection and the prediction of quantitative genetic parameters – all of which will underlie decisions on what strategy of selection will be most effective.

Pathology   A major goal of plant breeding is to increase productivity and quality by selecting superior genotypes. A limiting factor in economic production is the impact of pests and diseases. Therefore developing cultivars, which are resistant to detrimental pathogens, has been a major contributor to most cost effective production with reduced agrochemical inputs. Similarly, nematodes, insect pests and viruses can all have detrimental effects on yield and/or quality. Therefore plant breeders must also have knowledge of **nematology**, **entomology** and **virology**.

Weed Science   The response of a genotype to competition from weed populations will have an effect on the success of a new cultivar. Cultivars that have poor plant establishment, or lack subsequent competitive ability, are unlikely to be successful, particularly in systems where reduced, or no, herbicide applications are desirable, or their use is restricted. Similarly, in many cases genotypes respond differently, even to selective herbicides. Herbicide tolerance in new crops is looked upon favourably by many breeding groups, although cultivar tolerance to broad-spectrum herbicides can cause management difficulties in crop rotations.

Food Science   Increasing end-use quality is being identified as one of the major objectives of all crop breeding schemes. As most crop species are grown for either human or animal consumption, knowledge of food nutrition and other related subjects is important.

Biometry   Managing a plant breeding scheme has aspects that are no different from organizing a series of large experiments over many locations and years. To maximize the probability of success it is necessary to use an appropriate experimental approach at all stages of the breeding scheme. Plant breeding is continually described as 'a numbers game'. In many cases this is true, and successful breeding will result in vast data sets on which selection decisions are to be made. These decisions often have to be made during short periods, for instance between harvesting one crop and planting another. Therefore, plant breeders are required to be good **data managers.**

Agronomy   It is the aim of crop breeders to predict how newly identified genotypes will perform over a wide range of environments. This will require research into agronomic features that may relate to stress tolerance, such as heat, drought, moisture, salinity, and fertility. These experiments are essential in order that farmers (the primary customer) are provided with the optimal agronomic husbandry parameters, which will maximize genetic potential of the new variety.

Molecular Biology   Advances in molecular biological techniques are having an increasing role in modern plant breeding. Molecular markers are increasingly used by plant breeders to help select (indirectly and directly) for characters that are difficult to evaluate in the laboratory, or are time consuming, or expensive to determine accurately on a small plot scale. Genetic engineering and other tissue culture operations are becoming standard in many plant breeding schemes and it is likely that further advances will be made in the future. Knowledge of all these techniques and continued awareness of ongoing research will be necessary so that new procedures can be integrated into the breeding scheme where appropriate.

Production   The contributions that farmers, and other growers, have made to varietal development should never be underestimated. It also should be noted that growers are the first customers for plant breeding products. The probability of a new cultivar being successful will be maximized (or at least the probability of complete failure reduced) if growers and production systems are considered as major factors when designing breeding systems.

Management   There is a need to manage people, time and money. It has already been stated that plant breeding is a multidisciplinary science and this means being able to integrate and optimize people's effort to effectively use breeders' time. The length of most breeding programmes means that small proportional savings in time can be valuable and it hardly needs emphasizing that breeding needs to be cost effective and therefore the cost of the programme is always going to be important.

Communication   Most varietal development programmes consist of inputs from more than one scientist and so it is necessary that plant breeders are good **communicators**. Verbal and written communication of results and test reports will be a feature in all breeding schemes. Research publications and grant proposals are of major importance, particularly to public breeders, if credibility and funding is to be forthcoming. Finally, at least some plant breeders must be good at passing on the essential information about the subject to future plant breeders!

Information Technology The science underlying plant breeding is continually advancing, the agronomic practices are continually being upgraded, the end-users' choices change and the political context continually affects agriculture. This means that it is vital that breeders talk to these different groups of people but also use what ever technology is available to keep up to date on developments as they occur – or better before they occur!

Psychic The success of most plant breeding schemes is not realized before many years of breeding and trialling have been completed. It may be twelve or more years between the initial crosses and varietal release. Plant breeders therefore must be *crystal ball gazers* and try to predict what the general public and farming community will need in the future, what diseases will persist twelve years ahead and what quality characters will command the highest premiums.

In summary, therefore, successful plant breeders need to be familiar with a range of scientific disciplines and management areas. It is not, however, necessary to be an expert or indeed an authority in all of these. However, greater knowledge of the basic science underlying the techniques employed, and of the plant species concerned, in terms of the biology, genetics, history and pathology will increase the chances of a breeder succeeding in developing the type of cultivars most suitable for future exploitation.

EVOLUTION OF CROP SPECIES

Plant breeding consists of the creation and manipulation of genetic variation within a crop species, and selection of desirable recombinants from within that variation. The process is therefore an intensification of a natural process, which has been ongoing since plants first appeared on earth. As soon as humans started carrying out settled agriculture they effectively started plant breeding. In this section the main features of crop plant evolution will be covered briefly. The study of evolution is a vast and detailed subject in itself, and it will not be possible to cover more than an introduction to

it in this book. Emphasis will be on the areas that are most important from a plant breeding standpoint.

Knowledge of the evolution of a plant species can be invaluable in breeding new cultivars. Studies of evolution can provide knowledge of the past changes in the genetic structure of the plant, an indication of what advances have already been achieved or might be made in the future, and help to identify relatives of the domesticated plant which could be used in interspecific or intergeneric hybridization to increase genetic diversity or introduce desirable characters not available within existing crops.

Why did hunter gathers become farmers?

It is difficult to arrive at a firm understanding as to why humans became a race of farmers. Early humans are believed to have been foragers and later hunters. Why then did they become crop producers? Farming is believed to have started shortly after the last ice age. At that time there may have been a shortage of large animals for hunters to hunt due to extinctions. Indeed, little is known about the order of agricultural developments. Did man domesticate animals and then domesticate crops to feed these beasts, or were crops first domesticated, and from this the early farmers found that they could benefit from specifically growing sufficient food to feed livestock? The earliest farmers may also have been fishermen who tended not to travel continually and were more settled in one region. In this latter case, perhaps the first farmers were women who took care of the farming operation while the males fished and hunted locally. It may simply have been that some ancient people became tired of nomadic travel in search of food, became bored with living in tents and opted for a quiet life on the farm! The answers are not known, although it can often be interesting to postulate why this change occurred. One misconception about the switch from hunting-gathering to farming is that farming was easier. It has been shown that gathering food requires considerably less energy than cultivating and growing crops. In addition, skeletal remains show that the initial farmers were smaller framed and more sickly than their hunter-gatherer counterparts.

Irrespective of the reason which caused mankind to cultivate crops, few would question that the beginning of farming aligned with the beginning of what most of us would consider civilization. Farming created communities, community structure and economies, group activities, enhanced trade and monetary systems to name but a few. There is also little doubt that the total genetic change achieved by early farmers in moulding our modern crops has been far greater than that achieved by the scientific approaches that have been applied to plant breeding over the past century. Given that these early farmers were indeed cultivating crops, it is not surprising that they would propagate the most productive phenotypes, avoid the individuals with off-taste, and choose not to harvest those plants which were spiny. Even today among peasant farmers there is a general trend to select the *best plants* for re-sowing the next year's crops. Early farmers may have used relatively sophisticated plant breeding techniques as there is evidence that some native Americans have a long established understanding of maintaining pure line cultivars of maize by growing seed crops in isolation from their production fields.

What crops were involved? And when did this occur?

Today's world food production is dominated by small grain cereal crops, with world production of maize (*Zea Mays*), rice (*Oryza sativa*) and wheat (*Triticum* spp.); each being just under 600 million metric tones annually (Figure 1.1). Major root crops include potato (*Solanum tuberosum*), cassava (*Manihot esculenta*), and sweet potato (*Ipimiea batatas*). Oilseed crops are soybean oil (*Elaeis guineensis*), coconut palms (*Cocos nucifera*), and rapeseed (*Brassica napus*). World production of fruit and

vegetables are similar where tomato (*Lycopersico esculentum*), cabbage (*Brassica oleracea*) and onion (*Alliums* spp.) are leading vegetable crops, whilst orange (*Citrus sinesis*), apple (*Malus* spp.), grape (*Vitaceae* spp.) and banana (*Musa aceminata* and *M. balbisiana*) predominate amongst the fruits. Many of these modern day crops were amongst the first propagated in agriculture.

Many studies have been made to determine the date when man first cultivated particular crops. The accuracy of dating early plant tissue has improved over the past half century with the use of radio-carbon methods. It should, however, be noted that archaeological material which remains well preserved has not proved easy to find. Many of the most significant findings have been from areas of arid environments (e.g. the eastern Mediterranean and Near East, New Mexico and Peru). These arid regions favour the preservation of plant tissue over time, and not surprisingly, are the areas where most archaeological excavations have taken place. Conversely, there is a lower probability of finding well preserved plant remains in regions with wetter, and more humid, climates. Therefore, archaeological information may provide an interesting, but surely incomplete, picture.

A summary of the approximate time of domestication and centre of origin of the world's major crop species, and a few recent crop additions, is presented in Table 1.1. It should be reiterated that many crop species have more than one region of origin, and that archaeological information is continually being updated. This table is therefore very much an over-simplification of a vast and complex picture.

Some of the earliest recorded information which shows human domestication of plants, comes from the region in the Near East known as the 'Fertile Crescent'

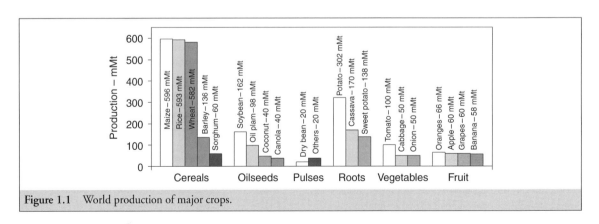

Figure 1.1   World production of major crops.

Table 1.1 Estimated time of domestication and centre of origin of major crop species. It should be noted that many crop species have had more than one suggested origin and that archaeological information is continually being updated. This table is therefore a simplification of an incompletely understood and complex picture.

Crop	Time of domestication (years)	Possible region of origin
Cereals		
Maize, *Zea Mays*	7000	Mexico, Central America
Rice, *Oryza sativa*	4500	Thailand, Southern China
Wheat, *Triticum* spp.	8500	Syria, Jordan, Israel, Iraq
Barley, *Hordeum vulgare*	9000	Syria, Jordan, Israel, Iraq
Sorghum, *Sorgum bicolour*	8000	Equatorial Africa
Oilseeds		
Soybean, *Glycine max*	2000	Northern China
Oil palm, *Elaeis guineensis*	9000	Central Africa
Coconut palm, *Cocos nucifera*	100	Southern Asia
Rapeseed, *Brassica napus*	500	Mediterranean Europe
Sunflower, *Helianthus*	3000	Western United States
Pulses		
Beans, *Phaseolus* spp.	7000	Central America, Mexico
Lentil, *Lens culinaris*	7000	Syria, Jordan, Israel, Iraq
Peas, *Pisum sativum*	9000	Syria, Jordan, Israel, Iraq
Root crops		
Potato, *Solanum tuberosum*	7000	Peru
Cassava, *Manihot esculenta*	5000	Brazil, Mexico
Sweet potato, *Ipimiea batatas*	6000	South Central America
Sugar beet, *Beta vulgaris*	300	Mediterranean Europe
Vegetables		
Tomato, *Lycopersico esculentum*	3000	Western South America
Cabbage, *Brassica oleracea*	3000	Mediterranean Europe
Onion, *Allium*	4500	Iran, Afghanistan, Pakistan
Fruit		
Orange, *Citrus sinesis*	9000	South-eastern Asia
Apple, *Malus* spp.	3000	Asia Minor, Central Asia
Grape, *Vitaceae* spp.	7000	Eastern Asia
Banana, *Musa aceminata, M. balbisiana*	4500	South-east Asia
Others		
Cotton, *Gossypium*	4500	Central America, Brazil
Coffee, *Coffea* spp.	500	West Ethiopia
Rubber, *Hevea brasiliensis*	200	Brazil, Bolivia, Paraguay
Alfalfa, *Medicago sativa*	4000	Iran, Northern Pakistan

(including the countries of Turkey, Syria, Israel, Iran and Iraq). Domestication of crops in this region surrounding the *Tigris River*, began before 6000–7000 BC. Two of the worlds leading cereal crops, einkorn and emmer wheat, and barley have their centre of origin in this region. In addition, archeological remains of onion, peas, and lentil, dating back to over 7000 years have all been found within the Fertile Crescent. In the Americas, similar or slightly later, dates of cultivation have been shown for beans and maize in central Mexico and

Peru, and potato, cassava, and sweet potato in Peru and western South America. Sunflower (*Helianthus*) is the only major crop species with a centre of origin in North America and indeed most other crops grown in the USA and Canada evolved from other continents. Rice, soybean, sugarcane (*Gramineae andropogoneae*), and the major fruit species (orange, apple and banana) were all first domesticated in China and the Asian continent a few millennia BC. Examination of archeological remains show that the dates of crop domestication in Africa were later; yet sorghum, oil palm, and coffee are major world crops that have their centre of origin in this continent. Similarly, cabbage and a few other vegetable crops have their centre of origin in Europe. Given more research, it may be found that many more of today's crops were domesticated at earlier periods.

Several crops of importance have been domesticated relatively recently. Sugar beet was not grown commercially in Europe until the 18th century, while rubber, date palm and coconut palms were not domesticated until the end of the 19th century. The forage grasses, clovers, and oilseed rape (*Brassica napus* L. or *B. rapa* L.) also are recently domesticated crops, although some researchers would argue that these crops have yet to make the transition necessary to be classified as truly domesticated. New crops are still being recognized today. The advent of bioenergy crops has identified the oilseed crop camelina (*Camelina* spp.), and the biomass crop switchgrass (*Panicum virgatum*) as potential new crops species which have yet to be grown in large scale commercialization.

A high proportion of today's major crops come from a very small sub-sample of possible plant species (Figure 1.1). It has been estimated that all the crop species grown today come from 38 families and 91 genera. Therefore, the source of our present day crops are more diverse than we have shown, although they still only represent a fraction of the total families and genera which have been estimated to exist within the angiosperms as a whole. Also, it should be noted that the sources of origin of these crops are spread over Europe, the Near East, Asia, Africa and America.

At some time in the past, each of our present day crop species must have originated in one, or more, specific regions of the world. Originally it was thought that there were only 12 major centres of origin including the Near East, Mediterranean, Afghanistan, the Pacific

Rim, China, Peru, Chile, Brazil/Paraguay and the USA. More recent research has altered this original view and it is now apparent that:

- Crops evolved in all regions of the world where farming was practiced.
- The centre of origin of any specific crop is not usually a clearly defined geographic region. Today's major crops are more likely to have evolved over large areas.
- Early farmers and nomadic travellers would have been responsible for widening the region where early crops have been found and added confusion concerning the true centres of origin.
- Regions of greatest crop productivity are rarely related to the crop's centre of origin.

Overall therefore, domesticated crops have originated from at least four of the six world continents (America, Europe, Africa and Asia). Australian aborigines remained hunter-gatherers and did not become farmers, and indeed farming in Australia is a relatively new activity started after western settlers arrived there. No surprisingly therefore few of today's major agricultural crops originated in Australia; however, a recently domesticated crop (Macadamia nuts) does have its origin in this continent.

NATURAL AND HUMAN SELECTION

All domesticated crops have been developed from wild, "weedy" ancestors. Early farmers modified weed species into modern-day crops through a process of genetic manipulation and selection. As a result these crop species have been sufficiently altered such that they can be considered to be domesticated. A definition of domestication has been given by Professor N.W. Simmonds as follows: "*a plant population has been domesticated when it has been substantially altered from the wild state and certainly when it has been so altered to be unable to survive in the wild*". The first part of this definition can certainly be readily accepted for almost all modern-day agricultural crops, although we still propagate many crops (e.g. date palm) where the crop species are modified only slightly from ancient ancestors. It is not always possible to relate domestication with a lack of potential to survive in non-cultivated situations since many commercially grown plants survive as volunteer weeds, or "escapes", in either the same, or different, regions to those in which they are most commonly grown commercially.

In the evolution of crop species we can often distinguish between natural and human selection. Natural selection tends to favour the predominance of the most adapted plant types, which manage to reproduce and disperse their progeny, while tolerating the stress factors that prevail in a particular environment. Therefore the natural selection favours plant phenotypes which have the greatest chance of survival, reproduction, and distribution of progeny. For example, wild cereal plants tend to have many small seeds at maturity and disperse their seed by shattering. These seeds also are likely to be attached to a strong awn to aid dispersal. Similarly, wild potato species produce many small tubers, have their tubers develop at the end of very long stolons (so that daughter plants do not have to occupy ground too close to the parent), and many have tubers with high levels of toxin, which discourage animals from eating them.

Human selection is the result of conscious decisions by a farmer or plant breeder to keep the progeny of a particular parent and discard others. Human selection is not usually directed to better survival in the wild (and indeed is often detrimental to survival outside cultivation). As an example, breeders have developed cereal cultivars which have fewer, but larger seeds, that do not shatter their seeds at maturity and that have a non-persistent awn. Similarly potato breeders have selected plants with fewer, but larger tubers, shorter stolons and with reduced levels of toxins in the tuber. Human selection also has produced crops that are more uniform in the expression of many of their characteristics. For example, they have selected seeds that all mature at the same time, with uniform germination, and fruits with uniform fruit size and shape. In more recent times plant breeders' selection has tended to result in shorter plants, greater harvest index, and increased ease of harvest. A large number of our crop species that used to require harvest by hand can now be harvested by machine, mainly as a consequence of their small stature and uniform ripening.

There is of course a range of characteristics that would have been positively selected both by natural evolution and early plant breeders. These might include aspects of yield potential, tolerance to stress factors and resistance to pests and diseases.

CONTRIBUTION OF MODERN PLANT BREEDERS

Around the turn of the 20th century the foundation of modern plant breeding was laid. Darwin's ideas on the differential survival of better adapted types were combined with those of Mendel on the genetic basis for the inheritance of plant characters. These two theories, combined with the research of scientists such as Weissman on the continuity of germplasm, and the analyses of Johannsen resulting in the idea of genotype/phenotype relationships, provided the scientific foundation of modern plant breeding.

There is little doubt that mankind has had a tremendous influence in moulding the morphology, plant types, end uses, and productivity of most crop species. Early farmers have taken wild, weedy plants and developed them into commercially viable agricultural crops. The contribution of modern plant breeding efforts is not always clearly defined nor can their achievements be easily measured.

Over the past century the world's human population has risen dramatically (Figure 1.2). World human population first exceeded one billion in 1804. It took a further 118 year of population expansion to double the world population. The human generation born after World War II (1945 to 1955) are often referred to as 'baby boomers'. Interestingly this is the first generation to witness the world's population double, from 3 billion to 6 billion individuals. It has further been estimated that within the next 20 years another 2 billion people will inhabit this earth.

Population explosion, combined with mass urbanization, and proportionally fewer farmers lead to fears from world population specialists of world-wide hunger and famine. However, since the start of the 'baby boom' era, the yield of almost all our major agricultural crops has increased as dramatically as human population. Cereal crop and oilseed crop production increased by over 120% and 130%, respectively (Figure 1.3). Similar increases in vegetable production of 80%, fruit production of 43%, pulses by 40% and root crops by 36% have taken place in a 50 year time span. When world

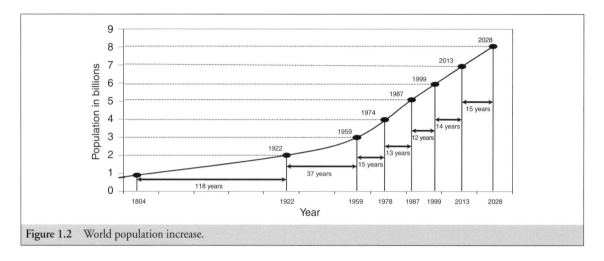

Figure 1.2   World population increase.

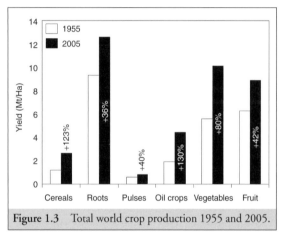

Figure 1.3   Total world crop production 1955 and 2005.

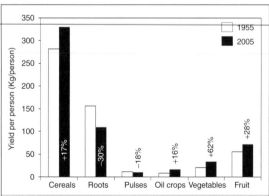

Figure 1.4   Total world crop production per capita 1955 and 2005.

agricultural production is adjusted according to population increase (Figure 1.4), cereal production per capita has increased by 17%, and fruit, vegetable, and oilseed production per capita has increased by 28%, 62% and 16%, respectively, while production of root crops and pulses has reduced per head of capita in the world.

These yield increases have been brought about by a combination of higher soil fertility (mainly due to additions of inorganic nitrogen fertilizers), improved chemical control of diseases and pests, better weed control through improved agronomic practices and herbicides, and better crop agronomic practices (e.g. correct plant densities) as well as by growing genetically improved cultivars.

So, how much of the improved yield can be attributed to the plant breeder (i.e. genetic change) and how much to better farming practices (i.e. environmental change)?

Yield increases of more than 100% have been found between single cross maize cultivars over the traditional homozygous varieties. Many researchers have attributed this increase to the heterotic advantage of single crosses over homozygous inbred lines, and therefore conclude that the contribution of plant breeding must be very high. However, a complication arises when comparing single cross hybrids, where selection has been aimed at maximum hybrid productivity, against inbred lines which have been chosen for their combining ability rather than their own performance *per se*.

It might be suggested that the question could only be answered properly by growing a range of old and new varieties under identical agricultural conditions. Since most modern cultivars are dependent on high levels of soil fertility and the application of herbicides, insecticides and fungicides, these would have to be used in the comparison trial. However, older cultivars were not grown under these conditions. Certainly, older cereal varieties tend to be taller than newer ones and are therefore more prone to lodging (flattening by wind or rain) when grown under conditions of high soil fertility. These considerations also show that cultivars are bred to best utilize the conditions under which they are to be cultivated. Nevertheless, several attempts to compare old and new cultivars have been undertaken in an attempt to determine the contribution of modern plant breeding to recent yield increases.

In one comparison carried out in the United Kingdom, winter wheats ranging in introduction date from 1908 to 1980 were simultaneously evaluated in field trials. In a similar experiment, spring barley cultivars ranging in introduction dates from 1880 to 1980 were compared. From the wheat cultivars available in the mid-1940s the grain yield from this study was about 5.7 t ha^{-1} but from the most recently introduced cultivars from 1980, yields were about 50% higher. There was a similar improvement in barley yield over the same period of about 30%. Therefore, considering these studies, breeding contributed about half to the more than doubled cereal yield between 1946 and 1980.

In contrast, a study carried out in potatoes, with cultivars with dates of introduction from 1900 to 1982, (Figure 1.5) found that modern plant breeding had been responsible for a very small contribution to the more than doubled potato yield in the United Kingdom. This study in potato may, in part, explain why 'Russet Burbank', introduced before 1900, still dominates potato production in the USA; while the cultivar 'Bintji',

introduced in 1910, remains a leading potato cultivar in the Netherlands.

In conclusion, modern day crops have shown significant yield increases over the past century. It would be wrong to suggest that the major contributor to this increase has simply been a direct result of plant breeding. Increases have rather resulted from a combination of plant breeding and improvements in crop husbandry. For example, the increased use of inorganic nitrogen fertilizer has greatly increased wheat (and other cereal) yield. However, this was allied with the introduction of semi-dwarf and dwarf wheat cultivars that allowed high nitrogen fertilizer application without detrimental crop lodging. Without the addition of nitrogen fertilizers would the dwarf wheat cultivars have been beneficial? Perhaps not. However, would high nitrogen fertilizer application have been possible without the introduction of dwarf wheat cultivars? It is difficult to know. The overall increase achieved to date has resulted in both genetic and non-genetic changes in agriculture.

In the future the same is likely to be true: that the next leap in crop productivity will result from a marked change in agronomic practice, plus the introduction of plant types that can best utilize this husbandry change. What changes will these be? It is impossible to know with any certainty. Recent moves to reduced tillage systems may be one option that could be considered and that would require specific cultivars to maximize performance under these situations.

Similarly, advances in recombinant DNA techniques may result in the development of crops with markedly different performances and adaptations to those available today. Introduction of these crop types may necessitate a major (or minor) change in crop husbandry to utilize the potential of these genetically modified crops.

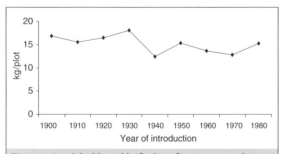

Figure 1.5 Saleable yield of tubers from potato cultivars grown over a three year period. Yield is related to the year that each cultivar was introduced into agriculture.

THINK QUESTIONS

(1) 'The yield of many crops species has risen dramatically over the last 50 years. This has been the direct result of plant breeding during this period and hence the trend is likely to continue over the next 50 years'. Briefly discuss this statement.

(2) Different crop species originated in different regions of the world. List the centre of origin of the following ten crop species: Onion

(*Allium*), Alfalfa (*Medicago sativa*), Rice (*Oryza sativa*), Potato (*Solanum tuberosum*), Soybean (*Glycine max*), Millet (*Eleusine coracana*), Cotton (*Gossypium* spp.), Sunflower (*Helianuthus* spp.), Wheat (*Triticum* spp.), and Apple (*Malus* spp.).

(3) 'The place of origin of crops, their history and evolution are events from the past and therefore have no relevance to modern plant breeding' True or False? Discuss your answer.

(4) Many believe that civilization (of man) started with the beginnings of agriculture. Basically there are two forms of agriculture: (1) rearing animals for meat, milk etc. and (2) raising crops for human or animal feed. No one knows which form of agriculture evolved first (or maybe both types started together). Explain why (in your opinion) one form came before the other or both forms evolved at the same time.

(5) A combination of natural selection, and selection directed by plant breeders (early and modern) has influenced the crops we now grow. List five charactereristics that mankind has selected which would not have been selected by a natural evolution process

(6) Have modern plant breeders improved the genetic fitness of our agricultural crop species, or have they simply selected plant types that are more suited to modern agricultural systems?

2

Modes of Reproduction and Types of Cultivar

INTRODUCTION

The most appropriate type of cultivar which can be developed to best fit the needs of a production situation will be determined, in part, by the breeding system and mode of reproduction of the species involved.

A cultivar (or variety) is defined as a group of one or more genotypes which have a combination of characters giving it **distinctness, uniformity** and **stability** (DUS).

Distinctness When a cultivar is 'released' for production it has to be proven that it is indeed new and that it is distinct from other already available cultivars. Distinctness is often defined on the basis of morphological characters that are known not to be greatly influenced by the environment. But other features such as physiology, disease or virus reaction, insect resistance and chemical quality may be used as well as, increasingly, molecular characterization in some countries (i.e. DNA markers). In the Guide to UK National Listings distinctness is described as follows: '*The variety, whatever the origin, artificial or natural, of the initial variation from which it has resulted shall be clearly distinguishable, by one or more important characteristics, from any other plant variety*'. This requirement is in part to ensure that new cultivars have not previously been registered by other breeding organizations. Information used to determine distinctness, also can be used later to identify and protect proprietary ownership of that cultivar so that other organizations cannot register the same cultivar or trade without permission, (within a set period) in that cultivar.

Uniformity Uniformity is related to the level and type of variation which is exhibited (usually phenotypic) between different plants within the cultivar. Any such variation should be predictable and capable of being described by the breeder. The variation should also be commercially acceptable and occur with no greater a frequency than that defined for that type of cultivar (as we will see below). The amount of variation that is permitted to exist in released cultivars varies according to the country of release. For example, in the United States, provided the degree and type of variation is clearly stated when the cultivar is released, the breeders can decide exactly how much heterogeneity exists for any character. In European countries, regulations regarding uniformity are more clearly defined and requirements for these to be adhered to. In the United Kingdom the guidelines read '*The plant variety shall be such that the plants of which it is composed are, apart from a very few aberrations, and account being taken of the distinctive features of the reproductive system of the plants, similar or genetically identical as regards the characteristics, taken as a whole, which are considered by the Ministers for the purpose of determining whether the variety is uniform or not*'.

Stability Stability of a cultivar means that it must remain true to its description when it is reproduced or propagated. Again the requirements for this differ between countries – in Europe it is generally by statute while in the United States and Canada it is usually considered to be the responsibility of the breeder to ensure stability. Again the UK guidelines give a description of what they mean by stability and it is '*The plant variety shall continue to exhibit its essential characteristics after successive reproductions or, where the breeder has defined a particular cycle of reproduction, at the end of each cycle of reproduction*'.

The requirement of distinctness is needed to protect proprietary cultivars and ensure that different organizations are not trying to claim the same cultivar and identify such cultivars as to their breeder. The requirements of uniformity and stability are there to protect the growers and to ensure that they are being sold something that will grow and exhibit the characteristics described by the breeder.

The further requirement of any new cultivar, is perhaps obvious, but nevertheless is a statutory requirement in many countries and referred to as **value** for **cultivation** and **use** (VCU). VCU can be determined by two primary methods and there will always be debate regarding which system is better. In the United Kingdom, which organizes statutory trials, VCU is described as follows: '*The quality of the plant variety shall in comparison with the qualities of other plant varieties in a national list, constitute either generally or as far as production in a specific area is concerned, a clear improvement either as regards crop farming or the use made of harvested crops or of products produced from these crops. The Qualities of the plant variety shall for this purpose be taken as a whole, and inferiority in respect of certain characteristics may be offset by other favourable characteristics*'.

In a few countries (including the United States) any plant breeder can sell seed from a cultivar developed and registered, irrespective of how well adapted it is to a given region or how productive the cultivar is likely to be. The choice of which cultivar to grow is left entirely to farmers and producers. It is common that farmers will allot a small proportion of the farm to plant a new cultivar, and if acceptable, will increase hectarage with time. Obviously, unadapted cultivars or those, which have inferior end-use quality, are unlikely to gain in acreage in this way. Similarly, companies (seed or breeding) and organizations rely on their reputation to sell their products and reputations can easily be tarnished by releasing and selling inferior products.

However, it is more common that countries have statutorily organized trialling schemes to determine VCU of cultivars that are to be released. This testing is usually conducted over two or three years, in a range of environments that the cultivars are likely to be grown. If breeding lines perform better than cultivars already available in that country, then government authorities will place that cultivar on the National List. Only cultivars that are included on the National List are eligible for propagation in that country. In some countries, newly listed cultivars also are entered into further statutory trials for one to two additional years. Based on performance in these extra years' trials, cultivars may be added to a *Recommended Varieties List*. This effectively means that the government authority or testing agency

is recommending that is would be a suitable new cultivar for farmers to consider. The theoretical advantage of statutory VCU testing is that it only allows 'the very best cultivars' to be grown and prevents unadapted cultivars from being sold to farmers. The major drawbacks of the *Regulatory Trialling* schemes are:

- Mistakes are inevitably made (although it is difficult to estimate at what rate) in that potentially desirable adapted new cultivars simply do not do well in the test conditions, for whatever reason. In this situation the cultivar is removed from further screening and all the time and effort expended by the breeder on that genotype's development would have been wasted.
- Authorities (or their agents), who organize these trials are often limited by resources and cannot always evaluate the number of test entries that may be submitted as thoroughly as might be desirable. In these situations there is often a 'lottery system' introduced where: companies are allowed to enter *a certain number* of test entries; not all entries are grown in all trial sites; and a non-comprehensive set of control cultivars included.
- Statutory trials suffer the same deficiencies as all small plot evaluation tests: they do not always reproduce or mimic the conditions or situations that occur on a realistic scale of production.
- They delay the period from a cultivar being developed to when it is released for commercial production. In some crop species (e.g. potato) this is not a great problem as the rate of increase of seed tubers in potato is low and it would normally take several years to increase sufficient tuber seeds to be planted on a commercial scale (a rate of increase of approximately 10:1). However, in other seed crops, for example rapeseed (oilseed rape), the rate of seed increase (particularly if off-season increase is possible) can be considerable, around 1000:1, and a three to four year delay in release can be costly to breeding companies.

The criteria for judging both DUS and VCU will be strongly determined by the type of species, particularly its mode of reproduction and multiplication for production.

MODES OF REPRODUCTION

It is essential to have an understanding of the mode(s) of reproduction prior to the onset of a plant breeding programme. The type of reproduction of the species (at least in commerce) will determine the way that breeding and selection processes can be maximized to best effect. There are two general types of plant reproduction **sexual** and **asexual**.

Sexual reproduction

Sexual reproduction involves fusion of male and female gametes that are derived either from two different parents or from a single parent. Sexual reproduction is, of course, reliant on the process of **meiosis**. This involves megaspores within the ovule of the pistil and the male microspores within the stamen. In a typical diploid species, meiosis involves reductional division by meiosis of the $2n$ female cell to form four haploid megaspores, by the process of *megasporogenesis*. This process in male cells, to form microspores, is called *microsporogenesis*. Fertilization of the haploid female cell by a haploid male pollen cell results in the formation of a diploid $2n$ embryo. The endosperm tissue of the seed can result from the union of two haploid nuclei from the female with another from the pollen, and hence ends up as being $3n$.

Asexual reproduction (**mitosis**) is the multiplication of plant parts or by the production of seeds (apomixis) that do not involve the union of male and female gametes. The process of mitosis will result in two cells that are identical in genetic make-up and of the same composition as the parental cell.

Seeds are effectively classified according to the source of pollen that is responsible for the fertilization. In the case of self-pollination the seeds are a result of fertilization of female egg cells by pollen from the same plant. Cross-pollination occurs when female egg cells are fertilized by pollen from a different plant, usually one that is genetically different. As a result plant species are usually classified into **self-pollinating** and **cross-pollinating** species. This is of course a gross generalization. There are species which are effectively 100% self-pollinating, those that are 100% cross-pollinating but there exists a whole range of species that cross-pollinate or self-pollinate to varying degrees. From the top 122 crop plants grown worldwide, 32 are mainly self-pollinating species, 70 are predominantly cross-pollinating, and the remaining 20 are cross-pollinating but do show a degree of tolerance to successive rounds of inbreeding.

The method of pollination will be an important factor in determining the type of cultivar that can, or

will, be most adapted to cultivation. For example, most species that can be readily used in hybrid production are generally cross-pollinating but need to be tolerant of inbreeding by selfing. This is because the hybrids are effectively the cross-pollinated progeny between two inbred genotypes.

Self-pollinating species are tolerant of inbreeding and consequently deleterious recessive genes are not common. They tend to have flower structures and behaviour that promote selfing. Individual lines of descent tend to approach homozygosity, shows little heterotic advantage when out-crossed and, individually, tend to have a narrower range of adaptation.

Cross-pollinating species tend to be intolerant to inbreeding, principally because they carry many deleterious recessive genes (these exist in the populations since they can be tolerated in heterozygous form). Generally, cross-pollinating species:

- Have a crossing mechanism that promotes outcrossing
- Show greater heterotic effect
- Are more widely adaptable to many different environments
- Have individual plants that are highly heterozygous at many loci

Particularly important are the outcrossing mechanisms. Cross-pollinating species often have distance barriers, time barriers or other mechanisms, which limit or prohibit self-pollination. Plants may be monoecious, where separate male and females flowers are located on different parts of the plant (e.g. maize) or indeed dioecious, where male and female flowers occur on different plants. Cross-pollination is also favoured in many cases where male pollen is shed at a time when the female stigma on the same plant is not receptive.

Another, more clearly defined sets of mechanisms are those termed as *self-incompatibility*. Self-incompatibility occurs when a plant, which has fully functional male and female parts, will not produce mature seed by self-pollination. There is a set of mechanisms that have naturally evolved to increase cross-pollination within plant species and hence promote heterozygosity. Adaptation to environmental conditions is greater if wider ranges of genotypes are produced in a progeny (i.e. the progeny shows greater genetic variation). Thus the chances of survival of at least some of the progeny will be enhanced, or at least the chance of extinction will be reduced.

There are a number of mechanisms than can determine self-incompatibility in higher plants:

- Pollen may fail to germinate on the stigma of the same plant flower
- Pollen tubes fail to develop down the style and hence do not reach the ovary
- Pollen tube growth is not directed towards the ovule and hence pollen tubes fail to enter the ovary
- A male gamete that enters the embryo sac may fail to unite with the egg cell
- Fertilized embryos resulting from self-pollination but these do not produce mature seeds

In several species (e.g. *Brassica* spp.) self-incompatibility can be overcome by bud-pollination, where pollen is applied to receptive stigmas of plants before the flowers open, as the self-incompatibility mechanism is not functional at this reproductive stage. Self-incompatibility is rarely complete and usually a small proportion of selfed seed can be produced under certain circumstances. For example, it has been found that environmental stress factors (particularly caused by applying salt solution to developing flowers) tends to increase the proportion of self-seed produced.

Asexual reproduction

Asexual reproduction in plants produces offspring that are genetically identical to the mother plant, and plants that are produced this way are called **clones**. Asexual reproduction can occur by two mechanisms: *reproduction through plant parts* that are not true botanical seeds and *reproduction through apomixis*.

Reproduction through plant parts

A number of different plant parts can be responsible for asexual reproduction. For example, the following are some of the possible organs that are reproductive propagules of plants:

- A **bulb** is a modified shoot consisting of a very much shortened stem enclosed by fleshy leaves (e.g. a tulip or an onion);

- A **corm** is a swollen stem base bearing buds in the axils of scale-like remains of leaves from the previous years growth (e.g. gladiolus);
- A **cutting** is an artificially detached part of a plant used as a means of vegetative propagation.
- A **rhizome** is an underground stem with buds in the axils of reduced leaves (e.g. mint or couch grass);
- A **stolon** is a horizontally growing stem that roots at nodes (e.g. strawberry runners);
- A **tuber** is a swollen stem that grows beneath the soil surface bearing buds (e.g. potato).

Reproduction by apomixis

Asexual propagation of plant seeds can occur in obligate and facultative apomicts. In obligate apomicts, the seed that is formed is asexually produced while in facultative apomicts, most seeds are asexually produced although sexual reproduction can occur.

Apomixis can arise by a number of mechanisms that differ according to which plant cells are responsible for producing an embryo (i.e. androgenesis from the sperm nucleus of a pollen grain; apospory, from somatic ovary cells; diplospory, from 2n megaspore mother cells; parthenogenesis from an egg cell without fertilization; and semigamy from sperm and egg cells independently without fusion). Apomixis can occur spontaneously, although in many cases pollination must occur (pseudogamy) if viable apomictic seeds are to be formed. Although the role of pseudogamy is not understood in most cases, pollination appears to stimulate embryo or endosperm development.

TYPES OF CULTIVAR

It may seem obvious that modes of reproduction determine the type of cultivar that is produced for exploitation. Cultivar types include pure-lines, hybrids, clones, open-pollinated populations, composite-crosses, synthetics and multilines. Obviously it would be difficult, if not impossible to develop a pure-line cultivar of a crop species like potato (*Solanum tuberosum*) as it is mainly reproduced vegetatively, and has many deleterious (or lethal) recessive alleles. Similarly, pea (*Pisum sativum*) is almost an obligate self-pollinator and so it would be difficult to develop hybrid pea, if nothing else seed production is likely to be expensive. A brief description of the different types of cultivar is presented below.

Pure-line cultivars

Pure-line cultivars are homozygous, or near-homozygous, lines. Pure-line cultivars can be produced most readily in naturally self-pollinating species (e.g. wheat, barley, pea, soybeans). But they can also be produced from species that we tend to consider as cross-pollinating ones (e.g. pure-line maize, gynoecious cucumber and onion). There is no universally agreed definition of what constitutes a pure-line cultivar, but it is generally accepted that it is normally one in which the line is homozygous for the vast majority of its loci (usually 90% or more).

The most common method used to develop pure-line cultivars from inbreeding species is to artificially hybridize two chosen (usually) homozygous parental lines, self the heterozygous first filial generation (F_1) to obtain F_2 seed, and continue selfing future generations, upto a point where the line is considered to be 'commercially true breeding' maybe the F_6 or F_7. At the same time it has been common to carry out recurrent phenotypic selection on the segregating population over each generation.

Open-pollinated cultivars

Open-pollinated cultivars are heterogeneous populations comprised of different plants, which are genetically non-identical. The component plants tend to have a high degree of heterozygosity. Open-pollinated cultivars are almost exclusively from cross-pollinating species. Plants within these populations have been selected to a standard that allows for variation in many traits but which shows 'sufficient' stability of expression in the characters of interest. Stability of these traits can be used to pass the DUS requirements necessary for cultivar release. Examples of open pollinated cultivars would include onions, rye and non-hybrid sweet corn.

In developing outbreeding cultivars the initial hybridization (the point at which the genetic diversity and variation is exposed) is usually between two outbreeding populations. In this case segregation is apparent at the F_1 generation. Desirable populations are identified and improved by increasing the frequency of desirable phenotypes within them.

Hybrid cultivars

Hybrid cultivars (single cross, three-way cross, and double cross hybrids) are very homogeneous and highly heterozygous. A true F_1 hybrid cannot be reproduced from seed of the hybrid generation because the progeny would segregate and result in a very non-uniform crop (although sometimes F_2 hybrid cultivars are sown).

Hybrid breeding is perhaps the most complex of the breeding methods. The process of cultivar development involves at least two stages. The first stage is to select desirable inbred lines from chosen out pollinated populations. These inbred selections are then used in test crosses to allow their comparison and assessment in relation to their general or specific combining ability. Superior parents are selected and these are then hybridized to produce seed of the hybrid cultivar. The parent lines are then maintained and used to continually reproduce the F_1 hybrids.

Clonal cultivars

Clonal cultivars are genetically uniform but tend to be highly heterozygous. Uniformity of plant types is maintained through vegetative rather than sexual reproduction. Cultivars are vegetatively propagated by asexual reproduction (clones) including cuttings, tubers, bulbs, rhizomes and grafts (e.g. potato, peaches, apples, chrysanthemums). A cultivar can also be classified as a clone if it is propagated through obligate apomixis (e.g. buffelgrass).

Clonal varietal development begins by either sexual hybridization of two parents (often clones) or the selfing of one of them to generate genetic variability through the normal process of sexual reproduction. Most of the parental lines will be highly heterozygous and segregation will begin at the F_1 stage. Desirable recombinants are selected from amongst the clonal propagules. Breeding lines are maintained and multiplied through vegetative reproduction and hence the genetic constitution of each selection remains '*fixed*'.

Synthetic cultivars

Inter-crossing a set number of seed lines generates a synthetic variety. In the simplest sense a 'first generation two-parent synthetic' is very similar to an F_1 hybrid. Synthetic lines can be derived from cross-pollinated lines or self-pollinating lines, although the latter instance is not common. Synthetic cultivars have a series of categories (Syn.1, Syn.2, ..., Syn.*n*) according to the number of open pollinated generations that have been grown since the synthetic line was generated. For example the first generation synthetics are classified as Syn.1, if this population is then selfed or out pollinated the next generation is classified as Syn.2 etc.

The use of synthetic cultivars has been most successful in cases where crop species show partial self-incompatibility (e.g. alfalfa). Examples of other crops where synthetic varieties have been released include rapeseed (*B. rapa* cultivar types), rye, pearl millet, broom grass and orchard grass.

Multiline cultivars

Multiline cultivars are mixtures or blends of a number of different cultivars or breeding lines. Each genotype in the mixture will be represented by at least 5% of the total seed lot. Many multilines are the result of developing near isogenic lines and using these to initiate the mix. These cultivars are usually self-pollinating species. A multiline is therefore not the same as a synthetic where the aim is to maintain heterozygosity by inter-crossing between the parent lines. Multilines became popular with the aim of increasing disease resistance by reducing the pressure for a pathogen to evolve/mutate to overcome the biological resistance. For example, near isogenic lines of barley, which differ in that each line has a different qualitative disease resistance, could be mixed to make a multiline. The main thought is to make the epidemiology of the pathogen such that it would be less likely to evolve virulence to all resistance genes in the mixture.

Composite-cross cultivars

Composite-cross populations are cultivars derived by inter-crossing two or more cultivars or breeding lines. These cultivar types have all tended to be inbreeding species (e.g. barley or lima beans). After the initial hybridizations have been carried out the composite-cross population is multiplied in a chosen environment such that the most adapted segregants will predominate and those less adapted to these conditions will occur at lower frequencies. A composite-cross population cultivar is therefore continually changing and can be considered (in a very loose sense) similar to old land races. Breeders' seed can never be maintained as the cultivar was originally released.

ANNUALS AND PERENNIALS

Plant species are categorized into annuals and perennials. World crop plants are fairly evenly distributed between annuals (approximately 70 species) and perennials (approximately 50 species). All major self-pollinating crop species are annuals while the greatest majority of cross-pollinating crops are perennials. Perennials pose greater difficulty in breeding than most annuals. Most perennials do not become reproductive within the first years of growth from seed. Most perennials are clonal cultivars and this can cause additional difficulties in maintaining disease free parental lines and breeding material. Winter annuals require vernalization of chill treatment before moving from vegetative to reproductive and can increase the time necessary for developing cultivars.

REPRODUCTIVE STERILITY

Female and male sterility has been identified in many crop species. Genetic and cytoplasmic male sterility have been identified in several plant species. Plant breeders in designing breeding programmes can utilize sterility and breeding schemes can be designed specifically to accommodate sterility, particularly in developing hybrid cultivars. Sterility also can pose problems and limits to the choice of parental cross combinations that are possible.

THINK QUESTIONS

(1) Complete the following table by assigning a **YES** or **NO** to each of the 16 cells.

	Pure-line cultivar	Out pollinated cultivar	Hybrid cultivar	Clonal cultivar
Would cultivars of only one single genotype be feasible?				
Is heterosis a major yield factor in resulting cultivars?				
Are resulting cultivars propagated by means of botanical seeds?				
Can the seed, or plant parts, of a cultivar be used for its own propagation?				

(2) Inbreeding and outbreeding species tend to have different characteristics. Explain factors that would determine if a given species should be classified as inbreeding or outbreeding.

(3) A number of different cultivar types are available in agriculture. Outline the major features of the following cultivar types: Hybrid cultivars; Inbred line cultivars; Clonal cultivars; Multiline cultivars; Open-pollinated cultivars and Synthetic cultivars.

(4) List two inbreeding crop species and two out-breeding crop species that have been utilized as (1) pure-line cultivars; (2) out-crossing population cultivars; (3) hybrid cultivars; (4) clonal cultivars and (5) synthetic cultivars.

(5) Describe the major features of the following types of apomixis: Diplospory; Semigamy; Parthenogenesis; Apospory.

(6) List five different plant parts that can be responsible for asexual reproduction.

3

Breeding Objectives

INTRODUCTION

The first exercise which must precede any of the breeding operations (and indeed a task that should be continually updated) is preparing a breeding plan or setting **breeding objectives**. Every breeding programme must have well defined objectives that are both economically and biologically feasible. Many new cultivars fail when they reach the agricultural practice. In some cases these failures are associated with the wrong economic objectives. Similarly many excellent new genotypes fail to become successful cultivars because of some unforeseen defect which was not considered important or was overlooked in the breeding scheme.

Objectives then are the first of the plant breeder's decisions. The breeder will have to decide on such considerations as:

- What political and economic factors are likely to be of greatest importance in future years?
- What criteria will be used to determine the yielding ability required of a new cultivar?
- What end-use quality characters are likely to be of greatest importance when the new releases are at a commercial stage?
- What diseases or pests are likely to be of greatest importance in future years?
- What type of agricultural system will the cultivar be developed for?

All these will need to be considered and extrapolated for a time that is likely to be 12 to 14 years from the onset of the breeding scheme. It should also be noted that politics, economics, yield, quality and plant resistance are not independent factors and that interactions

between all these factors are likely to have an effect on the breeding strategy. It is only after answering these questions that breeders will be able to ask:

- What type of cultivar should be developed?
- How many parents to include in the crossing scheme, which parents to include, how many crosses to examine, to examine two-way or three-way parent cross combinations and why?
- How should progeny progress through the breeding scheme (pedigree system, bulk system, etc.)?
- What characters are to be selected for or against in the breeding scheme and at what stage should selection for these characters take place?
- How to release the variety and promote its use in agriculture?

PEOPLE, POLITICAL AND ECONOMIC CRITERIA

The final users of almost all agricultural and horticultural crops are consumers who are increasingly removed from agricultural production systems. In 1863, the United States Department of Agriculture (USDA) was created and at that time 58% of the US population were actual farmers. Indeed only a few decades ago, the majority of people in the western world were directly involved in agricultural and food production. However, in 2006, less than 2% of the United States population were directly involved with agriculture. This past century therefore has resulted in a dramatic shift away from working on the land to living and working in cities. Agricultural output has, and continues, to increase almost annually despite fewer and fewer people working directly in agriculture than ever before.

The major consumers of agricultural products are therefore city dwellers who are remote from agricultural production but obviously have a large influence on the types of food that is desirable. In addition, these non-agriculturists have a tremendous influence on the way that agricultural products are grown and processed, and plant breeders would be foolish not to consider end-users' likes and dislikes when designing our future crops.

Despite an overall shortfall in world food supply, many developed countries have an over abundance of agricultural food products available and consumers have become used to spending a lower proportion of their total earnings on food than ever before. In addition, consumers have become highly interested in the way that food products are grown and processed. Many consumers are interested in eating '*healthy*' food, often grown without agro-chemicals and without subsequent chemical colourings or preservatives. Obviously breeding for cultivars that are resistant to diseases, which can be grown without application of pesticides fits the needs of these consumers. However, many consumers want what they call '*more natural*' food, and this has in recent years had a large impact on agriculture and inter-country trade in agricultural commodities, particularly with the advent of large-scale commercialization of genetically modified (engineered) organisms (GMO's).

The discovery of recombinant DNA techniques that allow the transfer and expression of a gene from one species, or organism, into another was a breakthrough in modern plant breeding and offers some enormous advances in crop development. However, the first GMO crops to be commercialized have not been well accepted by many consumers. The reasons for this are numerous but are related to '*plant breeders playing God*', and a general mistrust among consumers regarding the health and safety of these *novel* products. In addition there are concerns amongst many groups that there might be environmental risks associated with GMO crops and that transgenes will *escape* into the ecosystem as volunteers or by inter crossing with related weeds. Other issues obviously are involved in this barrier of acceptance including, free trade agreements, and monopolies of transformation technology by a few companies world-wide.

Nonetheless, how many plant breeders, 10 to 12 years ago, had the insight to foresee such resistance in acceptance of products, that many consider

being 'a significant advantage over proceeding cultivars'? Almost all of the first GMO crops have genetic advantages that are unseen by the final consumer. These crop products look and taste the same as traditional crops, in most cases require as many agro-chemicals in production, and finally, they are no cheaper for the end-user to buy! Perhaps, breeders of the first GMO releases should have given more attention to the end-users needs and released the first GMO crops that had *consumer advantages* rather than grower advantages. It would have been interesting to see what difference this might have made on the first commercialization of these crops. However, this is not possible, and including recombinant DNA technologies as tools in cultivar development is one issue that many breeders and breeding programmes are at present giving serious consideration. The whole GMO issue is, however, a good example of where cultivars were developed and became completely unacceptable based not on their genetic potential, agronomic adaptability, or end-use quality, but a general perception of unacceptability amongst the consumer.

Consumers, farmers or housewives, also have a tremendous impact on all aspects of everyday life, as they are usually involved in democratically electing politicians that govern the nations of the world.

Many people would consider it impossible to try and predict what is in the mind of a politician. Politicians and political policy will, however, continue to have a large determining feature in shaping agriculture. For example, there may be a very cheap and 'safe' chemical available that controls a certain disease in your crop of interest. With this in mind the breeding objectives may not include selection or screening for biological resistance to this disease. Several years into the breeding scheme, it may be decided that the use of this chemical is harmful to the environment and government policy responds by withdrawing the use of the chemical and suddenly the need for resistance to this disease in breeding lines is vital. As an example, it is likely that the majority of organophosphate based insecticides will not be labelled for application to many of the crops that today depend on them for successful production. It has further been suggested that over 80% of all agrochemicals used in the United States will not be labelled for use in agriculture.

Many soil fumigants are highly toxic, volatile chemicals that have adverse depletion effects on our atmosphere. As a result many governments world-wide have banned the use of soil fumigants such as methyl bromide. One possible alternative to using synthetic fumigation is Brassicaceae cover crops (plow-down crops) or high glucosinolate containing seed meal soil amendments. Glucosinolates, *per se*, are not toxic, but when mixed with water in the presence of myrosinase, they degrade into a number of toxic compounds including: isothiocyanates and ionic thiocyanate. Past research has shown that different Brassicaceae species produce different quantities and types of glucosinolate which have greater or lesser pesticidal activity to different pests. Insightful plant breeders have recognized the potential of biopesticides and have found that interspecific hybridization between different Brassicaceae species offers an opportunity to develop '*designer glucosinolate*' plants with specific pesticidal effects.

Political government agencies (i.e. the Food and Drug Administration in the United States) can greatly influence consumer acceptance and choice. One recent example of this relates to *trans* fats in cooking oils. Research has proven that hydrogenated vegetable oils which contain *trans* fats have adverse effects on human health when included in diets. Indeed, the Food and Drug Administration now requires *trans* fat content to be listed on all food products. Traditional vegetable oils like oilseed rape (canola) and soy, although relatively low in saturated fats, are usually hydrogenated to avoid off-flavours in high temperature frying and to increase shelf life of the oil products. Rancidity and off-flavours in vegetable oil are caused by high concentrations of polyunsaturated fats. This has greatly raised the consumer awareness of *trans* fats and as a result, there is now high demand for vegetable oils which have a low polyunsaturated fat content. Perceptive canola and soy breeders had anticipated *trans* fat labelling and had low polyunsaturated (high oleic acid) cultivars of canola and soy available to meet market needs. These low polyunsaturated fat cultivars produce oils that show higher thermal stability, lower levels of oxidation products, and increased shelf life with minimal hydrogenation.

Political pressure can also have an influence on the types of crop that are grown. Within several countries in the world, and also groups of countries (e.g. the European Union) the farming community are offered subsidies to grow certain crops. As a result, over-production can occur, which can affect the world price of the crop, and hence influence the economics of farming outside the subsidized regions. Similarly, if crop subsidies are reduced or stopped, this also can

Figure 3.1   Volkswagen Beetle ('Bio-Bug') powered by biodiesel produced from mustard oil.

have a similarly large but opposite affect on the economics and hence directly affect acreage of the crop in these other regions. Crop price is always driven by demand, greater demand resulting in higher price. This can give rise to increased acreage, leading to over production, which leads to reduced crop prices.

The United States, like many other western countries, has become increasingly dependent upon imported oil to satisfy energy demand. It is possible to substitute oils from fossil fuels with renewable agricultural products (Figure 3.1). Therefore, bio-ethanol and biodiesel fuel, lubricating oil, hydraulic oil and transmission oil, can all be derived from plants. At present the agricultural substitutes are still higher in cost than traditional fossil derived equivalents. However, in the future this may change either as a result of a change in fossil oil costs or in further breakthroughs in either increasing crop productivity or the process needed to obtain these substitutes from agriculture. Governments in several countries have mandated that liquid fuel should contain a certain proportion of biodiesel or bio-ethanol, and public transport vehicles in inner cities are being encouraged to use biofuels as these have fewer emission problems. Similarly, in many countries (particularly Northern Europe) governments are legislating that certain operations (e.g. chain saw lubricants) use only biodegradable oils. Also taxation decisions made by relevant countries can affect the relative cost and hence use of fossil and plant derived fuels and oils.

Economic criteria are important because the breeder must ensure that the characteristics of cultivars that are to be developed are the ones that will satisfy the farmers and end-users and that can be produced in an agricultural system at an economic level. The supply of a product and the consumer's demand for that product are inter related. If there is over-production of a crop then there is a tendency for the purchase price to be lower. Conversely of course, when a product is in high demand and there is limited production then the product is likely to command high premiums. In general, however there is a tendency for an equilibrium; that farmers will only produce a volume that they think they can sell according to the needs of the end-user.

Although as stated, plant breeding is influenced by economic activity, financial concerns are rarely considered in setting breeding objectives or setting a breeding or selection strategy. It is often assumed that increased yielding ability, better quality and greater disease or pest resistance are going to be associated with improved economics of the crop. Unfortunately this topic has been examined by very few researchers and is an area where greater examination would help the probability of success in plant breeding.

Private breeding companies have developed economic breeding objectives that have had greatest influence on breeding objectives or strategy. This is perhaps not surprising as these groups require developing better cultivars and selling seeds or collecting royalties on these genotypes in order to survive in the industry. Public sector breeding groups are often not under the same economic restraints. Thus, the private breeder in maximizing profits will tend to favour objectives that will promote seed sales and discourage farmers from retaining a proportion of their crop as seed for the following season. In many cases therefore hybrid cultivars are preferred over homozygous lines or open-pollinated populations. Similarly, private breeding companies have little incentive to develop varieties which have biological resistance to seed-borne diseases (e.g. virus disease in potatoes) as this will again encourage farmers to return each season to the seed company for new seed supply. In addition, many private breeding groups are linked to (or indeed owned by) agricultural chemical companies. There could therefore be reduced benefit in developing cultivars which are biologically resistant to diseases if that chemical company has a monopoly in sales of the chemical that is used to control the disease. Therefore the objectives of different groups will differ according to the organization that is funding the breeding programme.

INCREASING GROWER PROFITABILITY

In general yield is the most important character of interest in any plant breeding programme. Therefore, increasing crop yield will always be a sensible strategy. There would be only limited use for a new cultivar unless it has the potential to at least yield at a comparable level over existing varieties, unless the harvestable product fits a particular niche market and hence can attract a higher 'per unit price'. Plant breeders tend therefore to select for increased profitability. Farmer's profits are related to input costs and gross returns on the crop. Plant breeders can increase grower's profitability by:

- Increasing the yield per planted area, assuming input costs remain constant
- Increasing the region of crop production
- Reducing input costs while maintaining high yield per unit area (input costs will include herbicides, insecticides and fungicides)
- Increasing the inherent quality component of the end product so that growers receive a higher unit price when the harvestable product is sold, or such that the product is more nutritious

All crops have restricted ranges of environments to which they are adapted. Bananas and sugarcane are unlikely to be grown as commercial crops in the Pacific Northwest region of the United States or in Northern Europe. However, one attribute to increasing yield may be related to increasing the range of environments that a crop can be grown. For example, the development of earlier maturing *Brassica napus* lines has extended the canola (oilseed rape) acreage in Canada to include regions further west than was previously possible. A similar extension of adaptation must have been involved with movement of wheat and maize to northern temperate regions over the past decades. For example, potato production in many world regions is difficult, as healthy seed tubers cannot be produced or made available when required for planting. Developing potato cultivars that are propagated from true botanical potato seed (TPS) would overcome many difficulties that occur in these regions. Few potato diseases are transmitted through TPS. In addition, small quantities of TPS would be required for planting compared to traditional seed tubers, which are bulky, and usually require refrigerated storage.

These examples are, however, perhaps exceptions and it should be noted that the majority of breeding programmes are concerned with increasing yield potential within an already well established growing region.

Crop profitability is based on net profit rather than gross product. By reducing input costs and at the same time maintaining high yield per unit area, breeders can increase crop profits. Input costs in crop production include herbicides, insecticides and fungicides. Therefore developing cultivars, that are more competitive with weeds, resistant to damage by insect pest, or have disease resistance, reduce inputs and alleviates the need to purchase and apply chemicals. Other inputs will include nutrients (mainly nitrogen) and water (which can have a high price in irrigated farming regions). It follows, of course, that developing cultivars that require less nitrogen or are more tolerant to drought and other stress factors will result in greater profitability to growers.

Increasing harvestable yield

Yield, in the eyes of breeders, is considered to have two main components: *biomass*, the ability to produce and maintain an adequate quantity of vegetative material, and *partition*, the capacity to divert biomass to the desired product (seeds, fruits, or tubers etc.). Therefore, the partitioning of assimilate is very important in obtaining maximum yielding ability. Partition, in general, takes the form of enhancement of yield of desired parts of the plant product at the expense of unwanted plant parts (sometimes referred to as increasing the harvest index). This can take three main forms:

- Vegetative growth is reduced to a minimum compared to reproductive growth. In many crop species plant breeders have selected plants that have short stature, or indeed are dwarf mutants. This was mainly driven by the need to reduce lodging (plants falling over before harvest) under increasing levels of applied nitrogen. However, short plant stature also allows more convenient harvest and sometimes (e.g. in fruit trees) allows for mechanical harvest and hence avoiding more expensive harvest by hand-picking.

This strategy has been adopted in breeding objectives of many crops such as wheat, barley, oat, sunflower, several legumes, along with fruit trees like apple, orange, peach and cherry.

- Reproductive performance is suppressed in favour of a vegetative product. This has been applied to a number of vegetatively reproduced crops where sexual reproduction has been selected against in plant breeding programmes in favour of vegetative growth (e.g. potato, sugarcane, sugarbeet and various vegetables).

- Vegetative production to different vegetative parts can be used to increase yield of root and vegetable crops like potato, rudabaga (swede) and carrot. In this case, the breeder's task is to maximize the partition towards one type of vegetative yield (e.g. tubers in the case of potato) while maintaining the minimum biomass of unused plant parts (e.g. the haulm of potato).

Selection for yield increase

It is perhaps ironic that harvestable yield is arguably the most important factor in all plant breeding schemes and yet it is possibly the most difficult to select for. Increasing yield is complex and involves multiple modifications to the plants' morphology, physiology and biochemistry. Yield is, not surprisingly, quantitatively inherited, and highly modifiable by a wide range of environmental factors. Evaluating accurately the genotypic response to differing environments and genotype–environment interactions are the major limiting factors to maximizing selection response in plant breeding. Despite advances in molecular marker selection (mainly quantitative trait loci), increased yield is achieved by evaluating the phenotype of breeding lines under a wide range of rather atypical environments.

Yield potential will be one of few characters that is evaluated (or at least considered) at all stages of plant breeding programmes. Plant breeding schemes begin with many (often many thousands) genotypes on which selection is carried out over years and seasons until the 'best' cultivar is identified, stabilized and increased. Usually the size of plots used for field evaluation trials increases with increasing rounds of selection. On completion of the selection process, surviving breeding lines must have produced phenotypically high yield in small unreplicated plots (often a single plant), and a variety of increasing plot sizes associated with advancing generations in the selection scheme. Towards the most advanced stages, a few breeding lines may be grown in farm-scale tests.

Irrespective of the crop species involved, most plant breeders have been successful by selecting for yield *per se* of the plant part of importance, rather than by selecting of modified assimilate partition within a new crop cultivar, albeit that selection for the first resulted in a difference in the second. A number of different plant breeders have selected for yield components (i.e. attributes which contribute to total yield such as number of ears per plant, number of seeds per ear and seed weight in the case of cereals or number of tubers per plant and tuber weight in potato) rather than for yield itself. In most of these cases, however, there has been little achieved in respect of increasing overall yield. The reasons behind this failure are complex but one factor is related to the negative relationships between many yield components. Therefore positive selection for one component is counter-balanced by a negative response in another.

It is the actual yield obtained by the farmer that is clearly an important criterion, and therefore factors such as '*harvestability*' come to the fore. Mechanical harvesters now carry out many harvest operations. If a given genotype is not suited to mechanical harvest its usefulness can be greatly limited. Therefore characters such as plant lodging, precocious sprouting, seed shattering or fruit drop are all factors which will reduce the harvestable yield.

In many instances the uniformity of morphological characters (seed/tuber/fruit – size, shape, colour etc.) have a great effect on '*useable yield*'. Obviously, the end-user has a demand for a product which has a certain size, shape or colour and any deviation (either genetic or environmental) from this appearance will reduce useable yield. Similarly, if user product of a crop is prone to develop defects when processed, or in storage, this will affect useable yield as the defects will not meet the required legislative standards, or customer expectations, and will need to be culled out. A secondary factor regarding defective products is related to the cost that is incurred in having the defective fruits/tubers/seeds removed. Uniformity of yield is more difficult to evaluate than yield itself. Often it is not possible to evaluate product uniformity with any accuracy in small plot trails and therefore many potentially highly uniform breeding lines may be wrongfully discarded in the early stages of selection.

Research by crop physiologists has provided a great deal of knowledge regarding plant growth models for yield, and we have developed the ability to predict actual yield from a wide range of different physiological measurements. In the latter half of the last century, many plant breeders believed that input from crop physiologists and physiological biotype models of our crop plants would assist plant breeders to identify superior cultivars. Crop physiologists believed that photosynthetic or net assimilation rates could be used as selection tools to increase plant productivity and hence increase yield of crop plants. Some exceptions do exist and an example is afforded by lupins. Physiologically based research and modelling led to the proposal it would be beneficial to aid the development of the crop into Northern Europe to breed for a particular crop architecture, using genotypes with a determinate growth habit. Suitable mutants were found and indeed proved to be a marked improvement on the traditional lupin types in the new target environments. Despite the success in lupin, however, the impacts of physiological biotype models in plant breeding are rare.

INCREASING END-USE QUALITY

Irrespective of the yielding potential of a newly developed cultivar, success in agriculture will be determined by the end-use quality of the saleable product. Demand for sale of year-round fruits and vegetables has resulted in food products being shipped greater and greater distances to arrive fresh almost on a daily basis. In addition, greater emphasis is now, and will continue, to be placed on storage of perishable agricultural products to make them available at times of shortage of local supplies or, as noted above, to simply make them available on a year-round basis.

The are two main types of end-use quality:

- **Organoleptic** – consumer acceptance or preference of taste, size, texture and colour. Although many people differ in their preference, there is often agreement within taste panels as to general preference towards certain levels of expression of these attributes, thus 'liking' some genotypes over others (even disregarding 'off-tastes'). Similarly, the visual appearance of

Figure 3.2   Visual appearance of saleable products is an important characteristic for breeders of fruits and vegetables.

product (particularly with vegetables and pulses) can have a large influence on which are preferred over others (Figure 3.2). An increasingly important factor in today's agricultural products is their ability to be stored for long periods without loss of quality. Hence many vegetable or fruit crops have a certain '*harvest window*' where the majority of the crop is harvested. The end-user, however, demands that the product retain its quality characteristics.

- **Chemical** – where quality is determined by chemical analyses of the harvestable product. This is perhaps easily understandable in, say, oil crops, where the quality of the oil can be determined with great accuracy by determination of the oil fatty acid profile, or in the pharmaceutical (or *farm-aceutical*) industry where the quality of drugs are chemically determined. However, this is also true for fibre plants like cotton, and indeed fodder crops where protein content and digestibility are determined by analytical methods.

Several crop species have been utilized for a range of different uses according to variation in their physical or chemical characteristics. Take for example a potato crop. The end-uses of potatoes are either directly as raw tubers or through industrial processors, via retail purchasers. The needs and requirements of a potato will be different depending upon to the use that the product will be put to. For example, potatoes can be boiled, mashed, baked, chipped, canned, dried or fried. Each cooking method (or use) will demand certain quality characteristics. Boiled potatoes need to remain relatively firm and not disintegrate on boiling. This trait is related to the 'solids' content of the tubers, the lower proportions of solids being associated with less disintegration. Conversely, potato chip (crisp) processors do not wish to purchase potatoes with low solids as these have a higher water content, which has to be turned into steam (and hence waste) in the frying process. 'Chippers' also require potatoes with low reducing sugar content, which ensures that the chips (crisps) produced, will have a pale golden color.

Several crop species therefore have several end-uses, and specific quality characteristics will be required in cultivars designed for these uses. Bread wheat cultivars are required to have hard seed and high seed protein, while biscuit wheat has soft seed and low protein content. Edible rapeseed oil (canola) has a fatty acid composition low in erucic acid (22:1 fatty acid) and high in oleic acid (18:1), while industrial rapeseed cultivars need to have oil that is high in erucic acid content. The determination of most quality traits is based in their genes (i.e. is genetic) although the growing conditions of soil type, irrigation management and nitrogen application can all have large influences on the level at which these characters are expressed and hence on the final crop quality. Similarly, mechanical damage (particularly in vegetables and fruits) and crop disease can both greatly reduce the overall quality of the product irrespective of what the end use will be.

Determining '*desirable quality*' characteristics can be difficult and requires close integration of the breeding team with end-users and processors. Within some countries (e.g. the United States) government authorities have laid rules for quality standards (e.g. USDA #1 produce). In these cases it is often easier to set standards for the acceptable level of quality required from breeding lines. Caution however needs to be exercised since it is unlikely that these standards will remain constant through time, indeed they may change dramatically even before the new cultivar is even released!

Testing for end-use quality

If new cultivars are released which have special quality characters there may be justification, and economic merit, in introducing this as a '*specialty*' product even if overall yielding ability is not high. This would be justified if economic returns were sufficiently high to overcome the deficiencies in total yield. It should also be noted that competitors and other breeders will, of course, be quick to notice the market opportunity that has been opened and will focus on rapidly superseding such introductions, perhaps overcoming any of the obvious defects present in the original cultivar (e.g. pink grapefruit).

It is usually difficult (and most often impossible) to carry out an exactly similar processing operation on a large number of breeding lines. For example, in order to obtain the true quality potential of a new potato line with regards to French fry production (taking into account quality of end product, oil uptake, ease of processing, etc.), it would be necessary to produce several hundred tons of tubers and make French fries from them in a commercial processing plant. Similarly, in order to determine malting potential of barley for whisky (including all operations through to consumer

acceptance) would require large quantities of grain and considerable time. Obviously both these would be impossible in all but the very last stages of a breeding scheme. The basic features of any effective quality assessment in a plant breeding programme is that they should be **quick**, **cheap** and **use very little material**. These three criteria are important because:

- A plant breeding programme will involve screening many thousands of breeding lines each year or growing cycle.
- A plant breeder is often working against time. Many quality traits are assessed post-harvest and it is often important to make selection decisions quickly, before large-scale quality is determined.
- In most stages of a plant breeding scheme there are only limited amounts of material available for testing. It is often the propagative parts of the plants (seeds or tubers) that are used for testing and so a further complication is that many of the quality tests available are destructive of the very parts of the plant that are required to be grown to provide the next generation for selection.

It is clear therefore that it is important to determine at what stage in the breeding scheme to begin various quality screens. Obviously, quality evaluation should be included as early in the selection process as possible to avoid discarding some potentially high quality breeding lines. However, often this decision must be based on the cost of test, volume of material needed, accuracy of the test and the importance of the quality trait for the success of any new cultivar.

Taste panels (groups of experienced [or sometimes inexperienced] people who assess the food quality of new products) are often used. It is, however, impossible to compare more than a few types or breeding lines with a taste panel. These tests must also include some standard control lines, for comparative purposes, which further limits their potential when more than a few lines are to be tested.

Most other quality assessments are, at best, estimates of what will happen in the '*real world*'. They tend to be mini-reconstructions of parts of a larger scale commercial process or operation. When carrying out these assessments great care should be taken to ensure that the test follows as best as can be achieved the actual process that will happen in industry. It is therefore essential that good links are set up with industry personnel and that the breeding programme tries to integrate ideas from the processing industry into the breeding strategy as much as is feasible. This will also allow experiments to establish the levels of relationship (the correlations) between the 'lab tests' and the behaviour of the lines in commercial practice.

In other instances it is easier to record a related character than to record the trait itself. For example, in rapeseed breeding it is desirable to have low glucosinolate content in seed meal. Glucosinolate breakdown products are highly toxic and can cause dietary problems when seed meal is fed to livestock. Determining glucosinolate content is a two-day process and expensive. A much quicker and less expensive alternative is available. One of the breakdown products of glucosinolates is glucose. It is possible to obtain a good estimate of glucosinolate content simply by crushing a few seeds, adding water and estimating glucose concentration, using glucose sensitive paper. Similarly, malt barley breeders evaluate and select breeding lines for seed nitrogen content and soluble carbohydrates, which are highly related to the malting quality traits of malt extract and oligosaccharides, but these latter two characteristics are difficult to assess with small quantities of seed. Finally, the quality objectives of forage/fodder breeding programmes are biological in character and would ideally be met by testing the growth of animals fed on the breeding lines, but these large-scale feed studies are virtually never carried out, for reasons of time and cost. It should always be remembered, however, that these quality determinations are, at best, predictions, and in many cases, only a crude estimate of the character which is actually to be selected for.

Plant breeders seek to predict quality, however complex, by relatively simple measurements or organoleptic tests. Often small scale testing units based on the larger operation are used but in many cases quality assessment is determined by the correlation or relationship between an easily measured character and the more difficult to assess trait. However, before a new cultivar is released into agriculture it is desirable that new genotypes be tested on a commercial scale process. So wheats should be milled in a commercial mill, barleys should be malted and beer made, potatoes should be French fried and sold in fast food stores, onions stored in commercial storage and fodder fed to livestock before the product is released.

It is only after several rounds of testing at the commercial level that a secondary factor can be accurately estimated – that is the uniformity of quality. Uniformity of quality is as important as the actual quality character itself. A cultivar, which produces excellent quality in one environment or year but unacceptable quality in others, will have little merit in commercial production. Unfortunately, uniformity in quality (although one of the most important characters of a new cultivar) is difficult to assess within the restrictions of a feasible sized breeding scheme.

Overall, quality is what creates the demand for a product. It is the end-user who will mostly determine if that crop will be grown in future years. It is a very naive breeder who ignores the fact that consumer preference is continually changing and that the quality standards of today may be superseded by a new set of standards in the future. It is therefore imperative to organize a breeding scheme to be flexible and to try to cover as many potential aspects of yield, quality and other factors which may be important in the next two decades.

INCREASING PEST AND DISEASE RESISTANCE

A major limiting factor affecting both harvested yield and end-use quality of agricultural and horticultural crops is infection or infestation by plant pests and diseases. Breeding cultivars that are genetically resistant to pests and diseases is still a primary objective of plant breeding.

The development of resistant cultivars involves consideration of the genetic variability of the pest or disease as well as the variability in resistance (or tolerance) that exists within the crop species (or related species from which resistance can often be obtained). The durability of resistance of developed cultivars can be affected by the emergence of new races of the disease/pest that are able to overcome the resistance mechanism in the host plants. Thus the longevity of disease resistance that can be achieved in a new cultivar is often as important as the extent or degree of resistance that the new cultivar actually exhibits.

The major forms of disease and pests include fungi (air- and soil-borne), bacteria, viruses, eelworms and insects. However, this is not an exhaustive list. Other damage can occur (e.g. bird damage and mammal foraging) although in many cases it is difficult to imagine how biological plant resistance can greatly reduce such damage (cashew trees being knocked over by hippopotami is difficult to breed against!).

It can be assumed that pests and diseases will cause damage and almost all important diseases have been given attention by plant breeders. Crops differ greatly in the number of diseases that attack them and similarly in the exact damage that infection can cause. Small grain cereals are particularly susceptible to air-borne fungal epidemics; most solanaceous crops are especially affected by viruses, while cotton is particularly affected by insect attack.

It is difficult to assign importance to any class of plant diseases. In economic terms the soil and air-borne fungi may be more important than all other diseases. So much so that many breeding text books consider breeding for disease resistance, to actually be simply breeding for resistance to fungal disease. This is, of course, an over-generalization, and there is no doubt that other disease types also have potentially significant impacts on breeding objectives and goals, depending on the crop being bred. Indeed, it is recognized that virus diseases and many soil infestations are problematic because there are few treatments (especially agro-chemicals) that can be used to treat crops once plants become infected.

Any breeder trying to develop new cultivars with specific disease resistance must have knowledge of the particular disease or pest and its effect on the crop. One of the obvious, and most important effects of almost all crop pests and diseases, is reduction in yield. This is caused in four main ways:

- Destruction of leaf tissue and hence reducing plant photosynthesis capacity or efficiency (e.g. many rusts, mildew and blight).
- Stunting plants by metabolic disturbance, nutrient drain or root damage (e.g. many viruses, aphids or eelworm).
- Reducing plant stands by killing whole plants and leaving gaps in the crop which cannot be compensated with increased productivity of neighbours (e.g. vascular wilts, soil-born fungi and boring insects).
- Killing parts of plants (e.g. boring or feeding insects).

Killing plant tissue or causing reduced plant vigour can reduce yields *per se*, although reduced yield can result from other factors like increased weed infestation through reduced crop competition.

Other impacts of plant pests and diseases relate to damage to the end-use product of the crop. These infestations are usually initiated in the field but often become more apparent after harvest (e.g. cereal smuts, various rots and insect boring of fruits and tubers). Many pests/diseases also reduce the quality of harvested crops (e.g. insect damage in fruit or fungal blemishes of fruit or tubers).

The first task, which must often be carried out prior to screening for natural resistance to diseases, is to determine:

- Which diseases can affect the crop?
- What is the effect of these diseases on yield or quality?

Others who have been working with the crop in particular areas can often answer the first question. For example, if a relatively well established crop is to be bred (e.g. wheat in the Pacific Northwest of USA) there will already be a large body of data that has been collected regarding particular diseases and an indication of the frequency of disease attack.

The exact yield, quality or economic effect that different pests or diseases have on a crop can be used to partition the degree of effort that is exerted in breeding for resistance. Obviously, if a particular disease does not exist within the region there may be little point in devoting a large effort towards screening for natural resistance. Similarly, if a certain pathogen does not recognize your crop as a host, any attempt to increase resistance would be a waste of time and effort. In reality, the availability of a cheap and effective control measure will also decrease the priority a breeder assigns to tackling the resistance or tolerance to that particular disease or pest.

The most common means to determine the effect of a disease is to grow a series of genotypes under conditions where disease is artificially managed. In most cases the simplest way that this is done is to grow plots where disease is chemically controlled next to others where disease is allowed to occur naturally (or indeed artificially infected to ensure high disease pressure).

For example, the effect on yield caused by infestation by cabbage seedpod weevil (Figure 3.3) on four *Brassica* species was examined in field trials in 1992 and

Figure 3.3   Cabbage seedpod weevil larvae damage of canola seeds.

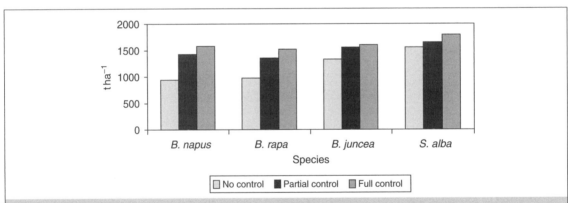

Figure 3.4   Seed yield (t ha^{-1}) from four *Brassicaceae* species as affected by late-season insect infestation when grown with full insect control, partial insect control and no insect control.

1993. Forty genotypes were grown in a pseudo-split-plot design where each entry was grown under three treatments: full weevil control with several insecticides, partial control with one insecticide, and no chemical control of the pest (Figure 3.4). The results differed between the four species investigated. However, without chemical control three of the four species showed yield reductions (some to a large degree). It is also obvious that *Sinapis alba* has more insect resistance (or tolerance) than the other species and indeed offered

breeders a source of resistance through intergeneric hybridization (see later). Additional data regarding cost of chemical application etc. can then be used in co-ordination with these data to estimate the actual economic effect of this pest on *Brassica* crop production.

A major difficulty in carrying out effective disease impact trials is to remove variation in as many other factors which may interact with those which are under study, as possible. For example, if the effect of a particular air-borne fungus is to be studied, then attempts must be made to ensure that other yield reducing factors (e.g. other air-borne fungi or other diseases) are kept under control in the trial.

Types of plant resistance

It has been claimed that for each gene in the host that controls resistance there is a gene in the pathogen that determines it, the pathogen will be *avirulent* (unable to overcome the resistance and hence unable to infect or injure the host) or *virulent* (able to infect or injure the host). This gene-for-gene hypothesis has been likened to a set of locks and keys. A simple example of this lock and key situation is shown in Table 3.1. In the first example, the host plant genotype has no resistance genes, so any pest genotype will be able to infect the host plant irrespective of the presence or absence of virulence alleles. In the second example, the host plant has one (or two) dominant genes for resistance to pest A strain (i.e. A_bb), and as the pest genotype has no virulence genes, the plant is resistant to the disease. In the third example, the host plant again has one (or two) dominant resistance genes against the A strain of pest but now the pest genotype is homozygous recessive (i.e. two copies) of the virulence gene ($a'a'$) and therefore unlocks (or can overcome) the resistance gene (A) in the host plant, and the plant is susceptible to the disease. In the last three examples the host plant has one or more copies of the dominant resistance genes to both pest A and B strains, (i.e. $A\_B\_$). When the pest genotype is homozygous for the recessive virulent alleles $a'a'$ but has no b' virulence

alleles (i.e. $a'a'B'B'$) the plant is resistant to the disease, as the pest opened (or overcame) the A resistance gene, but could not open the B resistance gene. Similarly, when the pest genotype is homozygous for the recessive virulent alleles $b'b'$ but has no a' virulence alleles (i.e. $A'A'b'b'$) the plant is also resistant to the disease, in this case because the pest opened (or overcame) the B resistance gene, but could not open the A resistance gene. In the last example, the pest genotype is homozygous recessive for both the $a'a'$ and $b'b'$ virulence alleles (i.e. $a'a'b'b'$) and can open (or overcome) both the A and B resistance genes in the host, and hence the host plant is susceptible to infection by the disease.

However, the situation is far from being this simple. Resistance to pests or diseases can be the result of either qualitative (single gene) or quantitative (multiple gene) inheritance. Resistance that is controlled by a single gene will result in distinct classes of resistance (usually resistant or susceptible) and are referred to as **specific** or **vertical resistance**. Resistance that is controlled by many genes will show a continually variable degree of resistance and is referred to as **non-specific, field, general** or **horizontal resistance**. Throughout this text the terms used will be vertical or horizontal resistance.

Vertical resistance is associated with the ability of single genes to control specific races of a disease or pest. The individual alleles of a major gene can be readily identified and transferred from one genotype to another. In many cases the source of the single gene resistance is derived from a wild or related species and backcrossing is the most common method to introduce the allele into a commercial background. Segregation of single genes can be predicted with a good degree of reliability and the selection of resistant genotypes can be relatively simply achieved by infection tests with specific pathogen races.

The primary disadvantage of vertical disease resistance is that new races of the pathogen are quite likely to arise that will be able to completely overcome the resistance. These new races may, in fact, have existed at a low level within the population of the pathogen before the resistance was even incorporated into the new cultivar or that cultivar was grown in agriculture. Thus, of course, the 'resistance' can be overcome relatively quickly. In addition, introduction of vertical resistance will increase the selective advantage of any mutant that arises in the pathogen population that can overcome the resistance. And as only a single mutation is required,

Table 3.1 Possible phenotypic plant response (i.e. resistant to disease or susceptible to disease) in various combinations of dominant alleles conferring single gene plant resistance (capital letters A or B represent resistance genes), and recessive alleles conferring susceptibility (a or b) and the 'matching' alleles in the pest where a′ or b′ confer virulence and A′ and B′ give avirulence.

Plant genotype	Pest genotype	Plant response
aabb	Any virulence gene	Susceptible
A_bb	No virulence gene	Resistant
A_bb	a′a′B′B′	Susceptible
A_B_	a′a′B′B′	Resistant
A_B_	A′a′b′b′	Resistant
A_B_	a′a′b′b′	Susceptible

the pathogen population may be many millions in size, so such a mutant will arise! New races of pathogen have overcome vertical resistance to air-borne diseases particularly quickly. In other cases, for example the single gene (H_1) in potato which gives vertical resistance to potato cyst nematode (*Globodera rostochiensis*), have proved very durable probably as a result of the much lower degree of mobility of the earth dwelling eelworm pest.

One technique used by plant breeders is to *pyramid* single gene resistance where there are a number of qualitative genes available. This technique was attempted in potato for late blight (*Phytophtra infestans*) using a series of single resistance genes (R genes) derived from *Solanum demissum*. To date, nine R genes have been identified, and up to six of these combined into a single potato clone. However, the late blight pathogen was able to overcome the pyramiding of R genes quickly and the technique was not successful. Pyramiding single gene resistance to diseases and pests, with the use of molecular markers that avoids the need to have suitable virulent pathotypes to screen for multiple resistance genes, has recently kindled interest.

Horizontal resistance is determined by many alleles acting collectively with each allele only having a small contribution to overall resistance. Because of the multiplicity of genes involved, horizontal resistance tends to be far more durable than vertical resistance. The advantage of horizontal resistance is in its ability to control a wide spectrum of races and, new races of the pathogen have difficulty overcoming the alleles at all loci controlling the resistance. The main disadvantage of horizontal resistance is that it is often difficult to transfer from parent to offspring. The probability of transferring all the resistant alleles from a resistant parent to a susceptible one can be very low. Breeding for horizontal resistance therefore tends to be a cyclic operation with the aim of increasing the frequency of desirable resistant genes.

Mechanisms for disease resistance

Two main disease resistance mechanisms exist. These are:

- Resistance due to lack of infection
- Resistance due to lack of subsequent growth or spread after initial infection

By far the most numerous examples of inhibition of infection in crop plants are related to hypersensitivity. Infection of the host plant causes a rapid localized reaction at the infection site. Host plant cells surrounding the infection point die, and hence the pathogen is effectively isolated from the live plant tissue and cannot spread further in the host plant. Hypersensitivity is usually associated with a necrotic flecking at the infection site and the host plant is totally immune to the pathogen as a result. Plant resistance through hypersensitivity is controlled by single genes and hence can usually be easily incorporated into breeding lines. In cases of high disease infection, cell death in the host plant can cause a significant reduction in plant photosynthesis and in extreme cases, plant death through lethal necrosis.

Other examples of disease infection inhibitors are less numerous and are usually associated with physical or morphological barriers. For example, the resistance to cabbage seedpod weevil found in yellow mustard (*Sinapis alba*) has been attributed to the very hairy surface of its pods and other parts, a feature not appreciated by the weevils that are deterred from laying their eggs. Similarly, leaf wax mutants of cabbage can deter insect feeding, and tightly wrapped corn husks can prevent insect pests from feeding on the developing seed.

Growth inhibition after infection is caused by the host plant restricting the development of a pathogen after initial infection. The pathogen is not able to reproduce in a resistant host plant as rapidly after infecting compared to a susceptible one. For example, Russian wheat aphids feeding on susceptible wheat plants inject toxins into leaves causing the leaves to fold. Adult Russian wheat aphids lay eggs in the folded leaves and the developing larvae gain protection from within the folds. Resistance genes have been identified that do not deter the adult Russian wheat aphids from feeding or injecting toxins into the leaves. However, the toxins do not cause the leaves of resistant wheat to fold, and hence there is greater mortality of developing Russian wheat aphid larvae, and reduced populations of the pest. Resistance to lack of spread of disease after infection can result from **antibiosis**, where the resistance reduces survival, growth, development or reproduction of the pathogens or insects feeding on the plant, or by **antixenosis**, where the resistant host plant has reduced preference or acceptance to the pest, usually insects. Resistance due to growth inhibition can be controlled by either qualitative or quantitative genes.

Another complication in screening and determining plant disease resistance is related to **tolerance**. Tolerance is the ability of a genotype to be infected by a disease and yet not have a marked reduction in productivity as a result. Therefore, despite plants showing disease symptoms (e.g. fungal lesions or insect damage) the plant compensates for the infection or damage. Tolerance to disease has been related to plant vigor, which may be associated with other physiological stresses. For example, genotypic tolerance in potato to infection by potato cyst nematode, late blight, early blight (Alternaria) and wilt (verticillium) are all highly correlated to drought stress or salinity tolerance.

One other factor needs to be considered in relation to disease resistance, and that is **escape**. This is where a genotype (although not having any resistance genes) is not affected by a disease because the infective agent of the disease is not present during the growth period of the genotype. Disease escape is most often related to maturity or other growth parameters of the plant and phenology of the pest. For example, potato cultivars that initiate tubers and mature early are unlikely to be affected by potato late blight (*Phytophthora infestans*) as the plants are mature before the disease normally reaches epidemic levels.

Testing plant resistance

In order to select for plant resistance to pests it is necessary to have a well established disease testing scheme, one that truly mimics the disease as it exists in an agricultural crop. If a plant's resistance to a pathogen cannot be reliably measured, it will not be possible to screen germplasm for differential resistance levels. Nor will it be possible to select resistant lines from amongst segregating populations.

Methods used for assessing disease resistance in plant breeding are extremely varied but may conveniently be grouped into three categories.

- Plants can be artificially infected in a greenhouse or laboratory. This can be especially effective in screening for vertical resistance (single gene resistance) where an all-or-nothing reaction is expected. Therefore a simple resistance or susceptible rating to

pathogens of the air-borne fungi is commonly detected on seedlings in the greenhouse or on detached leaves in the laboratory. Resistance to viruses is often carried out under greenhouse conditions where plants are 'hand-infected' with virus and plant response noted. Similarly, resistance to potato cyst nematode (both qualitative and quantitative resistance) can be effectively screened in a greenhouse by growing test plants in soil with high cyst counts. These methods in general demand rather precise pathological control but are usually quick, accurate and give clear results. If greenhouse or laboratory testing is to be used it is essential that the test results relate to actual resistance under field conditions. For example, seedling resistance to powdery mildew in barley is not always correlated with adult plant resistance observed under field conditions. The simplest method to authenticate small scale testing is to carry out direct comparisons with field trials as an initial part of setting up the testing regime.

- *In vitro* testing and screening has been used, where the diseases that infect plant tissue is a result of toxins produced by the pathogens. These toxins can be extracted and *in vitro* plantlets, or plant tissue, can be subjected to the toxin. As with testing in a greenhouse (above) it must be clearly shown that the *in vitro* plant tissue reaction to disease toxin is indeed related to whole plant resistance. It is never sufficient to observe phenotypic variation of *in vitro* material and to assume that this variation is useful *in vivo*. *In vitro* evaluation can be particularly effective if there are other reasons to propagate genotypes *in vitro*. For example if embryo rescue is necessary (as in some interspecific crosses), if whole plants are being regenerated from single cells (protoplast fusion or transformation), or if initial plant propagation is carried out *in vitro* to avoid disease (e.g. with potato).

- Field testing, using natural infection, artificial infection, or commonly, a combination of natural and artificial infection, is widely carried out. The object is usually to ensure that the initial infection rates are not limiting so that those estimates of genotypic resistance levels are more reliable (rather than simply a result of being escapes due to lack of infection). In the field there can be no precise control of the pathogen in terms of race composition or level of infection. Artificial infection may be achieved by inoculating seed or planting infecting material before planting the test plants, by inter-planting 'spreader rows' of plants that are highly susceptible to the disease, by spraying or otherwise infecting the crop with the disease or by artificial infection of the soil with eelworm or fungi. Uneven inoculation or infection is almost always a problem, especially with some of the soil pathogens. It is essential therefore, that these trials use adequate experimental designs and in many cases suitably high levels of replication.

In conclusion, breeding for disease resistance is no different (in many ways) to breeding and selecting for other traits. The steps, which must be taken in a disease breeding scheme, include:

- Develop a means to evaluate germplasm and breeding lines. In many cases success will be directly related to how effective the disease screening methods are at detecting differences in resistance levels. If it is not possible to differentiate consistently and accurately the level of disease present, it will not be possible to identify sources of plant resistance or to screen for resistance within segregating progeny.
- Evaluate germplasm and breeding lines to identify sources of plant resistance. In the first instance evaluation of the most adapted lines should be carried out. If no resistance is identified then more primitive or wild genotypes need to be screened.
- Examine, if possible the mechanism of resistance to determine the mode of resistance being exploited (e.g. avoid infection, limit spread, non-preference and antibiosis).
- Determine the mode of inheritance (i.e. qualitative or quantitative) of any resistance detected. The mode of inheritance can have a large influence on the method of introducing the resistance into the commercially acceptable gene pool (e.g. single gene would perhaps be best handled by a back-crossing method).
- Introgress source of resistance into a new cultivar. Despite the importance of disease resistance in plant breeding, it should always be remembered that a new cultivar will be unlikely to succeed simply because of disease resistance – new cultivars will also need to have acceptable levels of expression for the other important traits.

CONCLUSIONS

A house builder would not build a house without an architect first providing a plan, an automobile producer would not build an automobile without having some form of test model. So also a plant breeder will not produce a successful cultivar development programme without suitable breeding objectives. This will involve:

- Examining the whole production and use system from farmer, through processor to final product user. Determine the demands or preferences of each group in the production chain and take these into account in the selection scheme. Remember that there may be conflicts in the preferred requirements from different parts of this chain.
- Examine what is currently known about the crop. How is this end product produced at present? What cultivars presently predominate? What are the advantageous characters of these cultivars, and what are their defects?
- Examine how the operation of correcting present deficiencies can most effectively be addressed while always remembering that it is necessary to maintain at least the same level of acceptability for most other traits.

In order to set appropriate breeding objectives the breeder needs to consider incorporating: yield potential, disease and pest resistance, end-use quality and even the influence of potential political factors/decisions. Having taken these into account, it will be possible to design a successful plant breeding programme.

THINK QUESTIONS

(1) When breeding new cultivars it is often necessary to try and predict events that may occur in the future. Briefly outline four factors which may influence cultivar breeding and hence need to be considered in setting the breeding objectives of a cultivar development programme.
(2) Discuss the statement that durable resistance to disease is usually partial rather than complete.
(3) Outline the major features caused by the following diseases: soil-borne fungi; bacteria; eelworms; air-borne fungi; insects and viruses.

(4) Discuss the difference between durable and non-durable plant resistance. Indicate which of the disease forms listed in Think Question 3 (above) would be most likely to show durable qualitative disease resistance. Why?

(5) In plant breeding, two main disease resistance mechanisms exist, **inhibition of infection** and **inhibition of growth after infection**. Explain (with examples if necessary) each of the mechanisms.

(6) Sixteen members of an apple taste panel were asked to rate (according to taste) the fruit from each of three potential new apple cultivars (A, B and C). Rating scores were: 1 = best; 2 = second best and 3 = least best. The results are shown below:

Cultivar	Taste panel member															
	1	2	3	4	5	6	7	8	9	10	11	12	13	14	15	16
A	1	1	3	3	2	3	3	2	1	1	3	3	1	3	2	3
B	2	3	2	2	1	1	2	1	2	2	2	1	3	2	1	2
C	3	2	1	1	3	2	1	3	3	3	1	2	2	1	3	1

Without the use of second order (or higher) statistics, what are your conclusions?

In addition to an overall preference rating, list five questions which each taste panel member could have been asked either before or after tasting the samples.

(7) Choose any agricultural crop (e.g. wheat, barley, potato, apple, hops, etc.) and outline a set of breeding objectives to be used to develop new cultivars. Indicate potential markets for the new cultivars.

(8) Outline two political changes (either local, national or international) which would influence your decision when setting breeding objectives for a new potato cultivar that will be released in 18–20 years time.

(9) You have been offered a job as Senior Plant Breeder with the '*Dryeye Onion Company*' in Southern Idaho. Your first task in this new position is to set breeding objectives for onion cultivar development over the next 12 years. Outline the main points to be considered when setting your breeding objectives; indicate what questions you would like answered to enhance the breeding objectives you will set.

4

Breeding Schemes

INTRODUCTION

All successful breeding programmes have been designed around a breeding scheme. The breeding scheme determines the passage of breeding lines through the selection process, and through planting material increase for cultivar release. The process of selection will be carried out over a number of years, and under differing environmental conditions. The early selection stages of breeding programmes will involve screening many thousands of different genotypes. The early screening is therefore relatively crude, and in many instances involves only selection for single gene traits. After each round of selection, the '*better*', more adapted, or more disease resistant genotypes will be retained for further evaluation while the least adapted lines will be discarded. This process will be repeated over a number of years, at each stage the number of individual genotypes or populations is reduced and evaluation is conducted with greater precision in estimating the worth of each entry.

The breeding scheme used will be highly dependant on the crop species and the type of cultivar (inbred, hybrid, clone, synthetic, etc.) that is being developed. The general philosophy for developing a clonal cultivar like potato is therefore different from an inbred cereal cultivar, say barley. In the former, breeding selections are genetically fixed through vegetative propagation, but there will be a low rate of multiplication of planting materials. In the latter, there will be more rapid increase of planting material, although the segregating nature of the early generation breeding lines will complicate the selection process.

The most effective breeding schemes will utilize the positive attributes of a crop species while minimizing difficulties that might arise through the selection process. In the following section the breeding schemes for self-pollinating, out-pollinating, hybrid and clonal cultivars will be explained, along with mention on the schemes used for developing multilines and synthetics.

DEVELOPMENT OF SELF-POLLINATING CULTIVARS

Some crops that are generally produced as inbred cultivars are: barley, chickpea, flax, lentil, millet, peas, rice, soybean, tobacco, tomato and wheat.

One and a half centuries ago most inbred crop species were grown in agriculture as 'landraces'. Landraces were locally grown populations which were, in fact, a collection of many different genotypes grown in mixture and which were, of course, both genetically and phenotypically variable. Pure-line cultivars were developed first from these landraces by farmers who selected specific (presumably more productive) lines from the mixed populations and maintained these in isolation, selfed selections and eventually developed homozygous, or near-homozygous, lines. It is reasonable to assume that these homozygous lines were indeed more productive than the original landraces because by the end of the 19th century, landraces had almost completely disappeared in countries with an advanced agricultural system.

These early 'pure-line breeders' used the naturally existing genetic variation within the landraces they were propagating and probably the natural tendency of some species to self-pollinate (e.g. wheat) at a high frequency. However, this strategy has a limited potential and so modern plant breeders have to generate genetic variation and hence the three phase breeding schemes were established to **create genetic variation**, **identify desirable recombinant lines within the progeny** and **stabilize and increase the desired genotype**. It is interesting to note, however, that recently a number of plant breeders have returned to old landraces of wheat and barley to examine their wealth of genetic diversity as well as to testing combinations of lines in '*modern*' landrace combinations (i.e. multilines). Unfortunately most of the landraces that existed even 100 years ago are no longer available and potentially valuable germplasm and adapted combinations have been lost.

By far the most commonly used method of generating genetic variation within inbreeding lines is *via* sexual reproduction using artificial hybridization. There are, of course, other ways to produce variation. For example, variation can be produced by induced mutation, somatic variation, somatic hybridization and recombinant DNA techniques (all discussed in later chapters).

After producing the variation plant breeders will then traditionally screen the segregating population for desirable '*segregants*' while continuing to self successive generations, to produce homozygous lines. Thus, accomplishing the last two steps of the breeding scheme (selection and stabilization) more or less simultaneously.

Homozygosity

One of the difficulties in selecting desirable recombinant lines in pure-line breeding is related to segregating populations and the masking of adverse character expression as a result of the dominance/recessive nature of the segregating alleles in the heterozygotes. Another consideration is the relationship between genetical homozygosity and '*commercial inbred lines*'. The definition of complete homozygosity is that all the alleles at all loci are identical by descent, that is, there is not heterozygosity at **any** locus. However, for practical exploitation, the level of homozygosity does not need to be complete. Clearly the lines must basically breed '*true to type*' but this is by no means absolute. The degree of homozygosity can be directly related to the number of

selfing generations. Consider the simple case of just one locus A-a:

Parents		AA x aa		Frequency of heterozygotes
F_1		Aa		100%
F_2 Frequency	AA ¼	Aa ½	aa ¼	50%
F_3 Frequency	AA ¼	AA 1/8	Aa ¼ aa 1/8	aa ¼	25%

Consider just six of the loci that are involved, as set out below. Of these six loci two, only loci A and f have the same allele in both parents (which are both completely homozygous) and so the F_1 is homozygous at these two loci. At the other 4 loci the parents have different alleles and so the F_1 is heterozygous at these loci and these segregate in the F_2.

Parents	AAbbCCDDeeff × AABBccddEEff
F_1	AABbCcDdEeff
F_2	AABBCcDdeeff, AABbCcDDEeff, AAbbCCEeDdff, AABBccDdEeFF, etc

This can be generalized in mathematical terms as follows. Consider an F_1 that is heterozygous at n loci; heterozygosity (h) at any single loci after g generations ($g = 0$ at F_1) of selfing will be:

$$h = (1/2)^g$$

The probability (p) of homozygosity at n loci will be:

$$p = (1 - h)^n$$

Hence after g generations:

$$p = [1 - (1/2)^g]^n$$

This can also be written as:

$$p = [(2^g - 1)/2^g]^n$$

The level of homogeneity required in an inbred cultivar will depend to a varying extent on the personal choice of the breeder, seed regulatory agencies, farmers and end-users. Almost all pure-line breeding schemes involve selection of individual plants at one or more stages in the breeding scheme. The stages of single plant selection will have a large impact regarding the degree of heterogeneity in the end cultivar. If single plant selection is carried out at an early generation, say F_2, there may be greater heterogeneity within the resulting cultivar compared to a situation where single plant selection was delayed until a later generation, say F_8 where individual plants would be more homozygous. Breeders must ensure that a level of uniformity and stability exists throughout multiplication and into commercialization. Farmers will have preferences for cultivars which are homozygous, and hence homogeneous, for particular characters. These characters may be related to uniform maturity, plant height or other traits related to ease of harvest. Many believe that farmers are not concerned with uniformity of characters that do not interfere with end-use performance (e.g. flower color segregation). However, farmers take a natural pride in their farms and, therefore, like to grow '*nice looking*' crops and these are ones which are uniform for almost all visible characters. For most end-users there will be an obvious preference for cultivars which have high uniformity of desirable quality characters. For example, there may be a premium for uniform germination in malting barley or more uniform characters relating to bread making in wheat.

In contrast, some breeders of inbred cultivars like to maintain a relatively high degree of heterogeneity in their developed cultivars. They believe that this heterogeneity can help to '*buffer*' the cultivar against changes in environment and hence make the cultivar more *stable* over different environments. Often statutory authorities determine the degree of variability that is allowed in a cultivar. For example, all inbred cultivars released in the European Community countries, Canada or Australia must comply to set standards set for **distinctness uniformity** and **stability** (DUS) in Statutory National Variety Trials. In these cases it is common to have almost total homogeneity and homozygosity in released inbred cultivars.

With most breeders time is at a premium. Therefore, some methods are commonly used to reduce the time taken to achieve homozygosity and these include **single seed descent** and the use of **off-season sites** (this excludes the production of homozygous lines through doubled

haploids, which are relevant here but will be discussed separately in a later chapter).

Single seed descent

Single seed descent involves repeatedly growing a number of individuals from a segregating population, usually under high density, low fertility situations to accelerate seed-to-seed time. At maturity, a single seed from the self of every plant is replanted. This operation is repeated a number of times to obtain homozygous plants. Single seed descent is very well suited for rapid generation increase in a greenhouse where a number of growth cycles may be possible each year. Single seed descent in wheat and barley can be further accelerated by growing plants under stress conditions of high density, high light, restricted root growth, and low nutrient levels which result in stunted plants with only one or two seeds each but in a shortened growing period compared to growth under normal conditions (upto three or four generations in a year are possible in barley or canola).

It is very important, however, when using single seed descent, that unintentional selection is not being carried out for adverse characters. For example, in a single seed descent scheme in winter wheat (where plants will require a vernalization period prior to initiating a reproductive phase) vernalization requirements may be overcome artificially in a cold room. If this is done then care should be taken so that all seedlings do indeed receive sufficient cold treatment to overcome the vernalization requirement, otherwise the system will indeed be selecting these plant types with lower vernalization requirements. In addition, some genetic characteristics are not fully expressed when plants are grown under high competition stress conditions used for single seed descent. For example, the erectoides dwarfing gene (*ert*) is not expressed under single seed descent in the glasshouse, and therefore genotypes cannot be selected for this character under these conditions. In any case it is strongly advised that **no** selection be practiced during this phase.

Off-season sites

Off-season growing sites can also reduce the time for achieving a desired level of homozygosity. This is possible by having more than one growing season per year. Dual locations at similar latitudes in the Northern and

Southern Hemisphere are frequently used to increase either seed quantity or reduce heterozygosity in many breeding schemes. The use of off-season sites is often restricted to annual spring crops and there are only a few good examples where they have helped accelerate homozygosity in winter annuals and virtually none in breeding biennials.

If off-season sites are to be incorporated into the breeding scheme care must be taken to ensure that '*selectional adaptability*' of the off-season site does not have adverse effects on the segregating plant populations. For example the spring barley breeding scheme at the Scottish Plant Breeding Station used to increase F_4 breeding selections to F_5 by growing these lines over winter in New Zealand. Although New Zealand has a climate that is very similar to that found in Scotland there is a completely different spectrum of races of powdery mildew. As a result, mildew resistant selections made in New Zealand were of no relevance when grown in Scotland and so meant that all New Zealand trials needed careful spraying to avoid powdery mildew being confounded with other performance characters.

Breeding schemes for pure-line cultivars

There are probably as many different breeding schemes used by breeders of self-pollinating crops as there are breeders of inbreeding species. There are, however, three basic schemes: *bulk methods, pedigree methods* and *bulk/pedigree methods*. It should be noted that all the breeding schemes described would involve more than a single cross at the *crossing stage*. A number of these crosses will be two-parent crosses (Female parent × Male parent, say $P_1 \times P_2$), although many breeders use three- and four-way parent cross combinations ($[P_1 \times P_2] \times P_3$, and $[P_1 \times P_2] \times [P_3 \times P_4]$, respectively).

Bulk method

The outline of a bulk scheme is illustrated in Figure 4.1. In this scheme, genetic variation is created by artificial hybridization between chosen parents.

The F_1 and several subsequent generations, in the illustration upto and including the F_5 generations, are grown as bulk populations. No conscious selection is imposed on these generations and it is assumed that the genotypes most suited to the environment in which the bulk populations are grown will leave more offspring

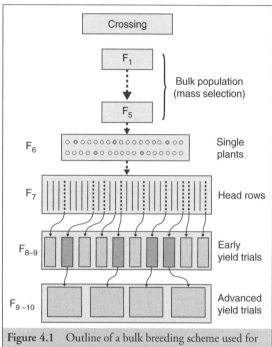

Figure 4.1 Outline of a bulk breeding scheme used for breeding inbreeding crop species.

and hence predominate in future generations. Similarly, these bulk populations are usually grown under the stress and disease pressures common to the cultivated crop, and it is assumed that the frequency of adapted genotypes in the population increases. It is therefore very important that the bulks are grown in a suitable and representative environment. After a number of rounds of bulk increase, individual plants showing desirable characteristics are selected at the F_6 stage. From each selected plant, a plant (or head) row is grown and the produce from the best lines/rows are bulk harvested, for initial yield trials. More advanced yield trials are grown from bulk harvest of desirable populations.

The major advantage of the bulk method is that conscious selection is not attempted until plants have been selfed for a number of generations and hence plants are nearly homozygous. This avoids the difficulty of selection among segregating populations where phenotypic expression will be greatly affected by levels of dominance in the heterozygotes. This method is also one of the least expensive methods of producing populations of inbred lines. The disadvantage of this scheme is the time from initial crossing until yield trials are grown. In addition, it has often been found that the natural selection, which occurs through bulk population growth, is not always that which is favourable for growth in agricultural practice. In addition, natural selection can, of course, only be effective in environments where the character is expressed. This often prevents the use of bulk methods at off-season sites.

Other methods have been used to produce homozygosity in bulk breeding schemes. These include single seed descent and doubled haploidy. Breeding schemes that use these techniques, have increased the popularity of bulk breeding scheme in recent years, as the time from crossing to evaluation can be minimized. However, the basic philosophy is similar, being to produce near-homozygous lines, and thereafter select amongst those inbred lines. Where rapid acceleration to homozygosity techniques are used, it is essential to ensure that no negative selection occurs. For example, research has shown that creating homozygous breeding lines of rapeseed (*B. napus*) through embryogenesis produces a higher than random, frequency of plants with low erucic acid in the seed oil. If low erucic acid content is desired, this poses a selection advantage. If, however, an industrial oilseed cultivar were desired (one with high erucic acid content) then using embryogenesis would be detrimental.

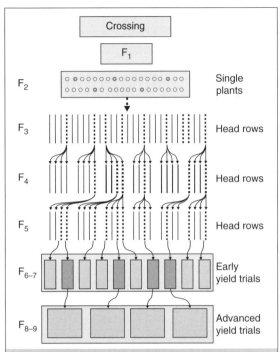

Figure 4.2 Outline of a pedigree breeding scheme used for breeding inbreeding crop species.

Pedigree method

The outline of a pedigree breeding scheme is shown in Figure 4.2. In a pedigree breeding scheme, single plant selection is carried out at the F_2 through to the F_6 generations. The scheme begins by hybridization between chosen homozygous parental lines, and segregating F_2 populations are obtained by selfing the heterozygous F_1s. Single plants are selected from amongst the segregating F_2 population. The produce from these selected plants is grown in plant/head rows at the F_3 generation. A number of the 'most desirable' single plants (in Figure 4.2 four plants) are selected from the 'better' plant rows and these are grown in plant rows again at the F_4 stage. This process of single plant/head selection is repeated until plants are 'near' homozygous (i.e. F_6). At this stage the most productive rows are bulk harvested and used as seed source for initial yield trials at F_7.

In addition to being laborious (as a considerable amount of record keeping is required) and relatively expensive, annual discarding may lead to the loss of valuable genotypes, particularly under the changing

environmental conditions from year to year, make selection difficult. Other disadvantages of the pedigree method are that it requires more land and labor than other methods; experienced staff with a '*good breeders' eye*' are necessary to make plant selections; selection is carried out on single plants where errors of observation are very large while actual yield testing is not possible in the early generations.

If selection was effective on a single plant basis, pedigree breeding schemes would allow inferior genotypes to be discarded early in the breeding scheme, without the need of tested in more extensive, and costly, yield trials. Unfortunately, pedigree breeding schemes offer little opportunity to select for quantitatively inherited characters and even single gene traits can cause problems when selecting on a single plant basis in highly heterozygous populations.

Bulk/pedigree method

The outline of a bulk/pedigree breeding scheme is illustrated in Figure 4.3. This type of breeding scheme uses a combination of bulk population and single

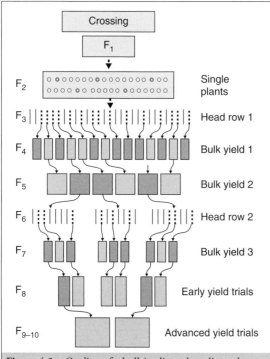

Figure 4.3 Outline of a bulk/pedigree breeding scheme used for breeding inbreeding crop species.

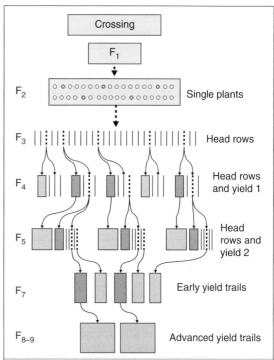

Figure 4.4 Outline of a modified bulk/pedigree breeding scheme used for breeding inbreeding crop species.

plant selection. An F_2 population is produced by selfing F_1 plants from artificial hybridizations. Individual plant selections from the segregating F_2 populations are grown in plant progeny rows at the F_3 stage. Selected F_3 populations are bulk harvested and preliminary yield trials are grown at the F_4 stage by planting the bulked F_3 seed. F_5 and F_6 bulk seed yield trials are grown, in each case by planting bulked seed from the previous year's trial. Selection of populations is based on performance in these trials. At the F_6 stage single plant selections are once again made from the now near-homozygous lines. Progeny from these plant selections are grown then as plant rows at F_7; second cycle initial yield trials at F_8 and more advanced yield trials at F_9.

The advantage of this combined breeding scheme is that inferior individuals, lines or populations are identified and discarded early in the breeding scheme. More than a single cultivar may be derived from a population or heterogeneous line identified as being superior by the earlier generation testing. Disadvantages will include, with fixed resources, the use of testing facilities

for evaluation of individual lines in the early generations and so reducing the number of more highly inbred lines that can be evaluated. Despite these disadvantages, bulk/pedigree schemes (or close derivatives of) are most commonly used to develop inbred cultivars.

Modified pedigree method

Most breeding schemes have developed breeding schemes that are combinations of bulk and pedigree methods. For example the breeding scheme used for developing winter barley cultivars at the Scottish Plant Breeding Station was a *modified pedigree trial scheme* (illustrated in Figure 4.4).

The modified pedigree trial breeding scheme enables yield trials to be grown simultaneously to pedigree selection. Single plants are selected from amongst segregating F_2 populations. Seed from these selections are grown as plant progeny rows at F_3. One, or more, single plants are selected from each of the desirable F_3 plant rows, and the remainder of the row is bulk harvested. The single plant selections are grown as plant progeny rows at F_4. Simultaneously, the harvested F_3 bulk is planted in a preliminary yield trial. The seed from the bulk yield trial is used to plant a more extensive bulk yield evaluation trial at the F_5 stage. Based on the results from the F_4 bulk yield trial, the most productive populations are identified. Single plant selections are selected from the corresponding plant progeny rows and the remaining row is bulk harvested for a further yield trial the following year at the F_5 generation. This process is repeated at the F_5 to F_6 stage, by which time near homozygosity is achieved in the remaining lines.

The advantage of the modified pedigree scheme is that it attempts to utilize progeny bulk evaluation for yield and other quantitatively inherited characters, while single gene traits can be screened for on a single plant/single row basis. In addition, this scheme allows for evaluation of quantitative characters while simultaneously inbreeding selections.

Number of segregating populations and selections

There have been numerous debates amongst plant breeders concerning the question of how many plants or populations should be evaluated, and selected, at each stage in a breeding scheme. Unfortunately, there is no simple recipe to help new breeders, and the questions can only be addressed from an empirical standpoint. Plant breeding is a *numbers game* and greater success is often associated with screening many thousands of breeding lines. However, plant breeding programs should only be as large as the specific breeding group can handle. Therefore, it is not productive to assess more lines at any stage in a breeding scheme than can be *effectively and accurately assessed*.

It is often easier to work backwards and ask how many lines you can handle at say, the advanced yield trial stage in the breeding scheme, and then move backwards to the previous stage and predict how many lines are required at that stage to ensure that the required number are selected, and so on.

Similarly, the number of initial cross combinations that should be used, differs markedly in different breeding programs. Often, a large number of crosses need to be screened, as the breeder cannot identify the most productive cross combinations. With experience of specific parents in cross combination and the benefit of

'cross prediction' techniques (see Chapter 7), it is possible to reduce the number of crosses screened on a large scale and hence allowing more effort to be spent on evaluating larger sized populations of the selected crosses.

Seed increases for cultivar release

At the other end of the breeding programme, once desired cultivars have been identified it is necessary to produce a suitable quantity of seed that will be grown and increased for varietal release. This seed lot is usually called *Breeders' seed* as in most cases producing this seed is the responsibility of the breeder rather than of a seedsman. It is vital that breeders produce breeders' seed lots that are pure, free from variants, and the genotype that is true-to-type to the cultivar that is to be released. Breeders' seed is used to plant '*foundation*' seed, which, in turn, is used to plant the various levels of '*registered*' or '*certified seed*', which is eventually sold to farmers.

Producing high quality breeders' seed is very similar to the breeding schemes (described above). In general there are two basic methods of producing breeders' seed **mass bulk increase** and **progeny test increase**.

Mass bulk increase

In mass bulk increase schemes a uniform sample of seed from the selected line is chosen and planted once to result in the breeders' seed lot. The advantage of this simple method is that it is inexpensive and takes only a single year to obtain the required seed. Disadvantages are mainly related to the purity, homozygosity and homogeneity of the cultivar entering into commercialization.

Progeny test increase

The progeny test increase method is more expensive and takes longer to obtain the seed required. This method is very similar to the bulk/pedigree breeding scheme. A number of single plants are selected from the homozygous/near homozygous advanced breeding line. These are grown as plant progeny rows. Individual plant rows are discarded according to off types or non-uniformity. The remaining rows are harvested individually and the seed from each row is used to plant larger progeny plots the following season. These plots are again inspected and those that do not have the required homogeneity or show off-types are discarded. At harvest the progeny plots are bulk harvested as breeders' seed. Breeders who wish to maintain a degree of heterozygosity in the released cultivar will include greater numbers of initial single plant selections in the scheme or they may not be as restrictive in decisions to discard progeny rows or plots. The advantage of the progeny multiplication method is that it allows greater control by the breeder and results in greater homogeneity in the released cultivar.

DEVELOPING MULTILINE CULTIVARS

Multiline cultivars (multilines) are mixtures of a number of different genetic lines (or indeed different species). Multiline cultivars are almost exclusively comprised of mixtures of lines from inbreeding species. Multilines have been developed for a number of different crop species including barley, wheat, oats, and peanuts. In turfgrass, intraspecific and interspecific multilines cultivars are grown commercially.

Multilines have been suggested as one means to minimize yield or quality losses due to diseases or pests that have multiple races and where the race specificities can change from year to year. Therefore there is a lower probability that all plants within the mixture (each with a specific disease resistance gene or resistance mechanism) would be affected as severely, over a period of years, as a homozygous cultivar. It has also been suggested that the use of multilines would result in more durable mechanisms of disease resistance in crop species.

Research has also suggested that multiline cultivars are more stable over a range of different environments than are pure-line cultivars. The reason for this has been related to the heterogeneous nature of the mixture where some lines in the mix do well in some years or locations while others perform better under different conditions. Therefore multilines show less genotype by environment interactions, a primary reason for their popularity with peanut breeders. Similarly, such considerations have led to mixtures of rye grass and Kentucky bluegrass being sold commercially. Rye grass has rapid emergence and establishment and does better than bluegrass in shaded areas.

Marketing of multiline varieties in the United States has advantages over other types of cultivar as seed companies can sell the seed without a common brand name if the seed sold is labelled with a '*cultivar not stated*' label. Multilines can also be sold under more than one name. For example, the same multiline can be sold with the brand names 'Browns Appeal', 'Browns Wonder' and 'Wonder Why' by the same or different seed sales groups. In other countries, however, multiline cultivars must comply with the set standards of DUS required for other inbred cultivars, and this has limited their use because of the difficulties in obtaining such homogeneity standards in a mixture.

The same care needs to be taken when producing breeders' seed from a multiline cultivar as is the case with a pure-line cultivar. The individual lines forming the mixture are increased independently by either mass or progeny multiplication methods (above). The individual components are then mixed in the proportions required, the seed mixed to form the breeders' seed, from which foundation seed is produced. The prevalent diseases, yielding ability or other appropriate factors will determine the proportion of lines within the mix. It is important when calculating multiline mixture proportions to take into account the seed size (if mixing by weight) and also the germination potential of each line (which may be different for the different lines).

One major complication relating to seed mixture proportions is the reproductive potential or productivity of each genotype in the mixture. For example, if the given proportion of two-parent lines (A and B) in a very simple multiline is 1 : 1 but the reproductive potential of A is twice that of B a 1 : 1 mix of breeders' seed will result in a 2 : 1 ratio of the lines being harvested from foundation seed and a 4 : 1 ratio being sold to the farmer after one further multiplication, to certified seed. Similarly, other environmental conditions may affect the proportions of mixed lines in the multiline. These changes could be related to foundation and certified seed being produced in an atypical environment or where a different disease spectrum exists.

Some multiline cultivars are mixtures of isogenic (or near isogenic) lines which differ for a single gene (usually conferring resistance to a certain strain of a pathogen). The most common method used in developing isogenic lines (lines which only differ in their genotypes by specified genes) in plant breeding is through the use of backcrossing. The genetics of **backcrossing** will be covered later in the qualitative genetics section. However a brief description will be given here.

Backcrossing

Backcrossing is a commonly used technique in developing pure-line cultivars. This technique has been used in plant breeding (not only in inbred species) to transfer a small, valuable, portion (usually a single gene) of the genome from a wild or unadapted genotype into the background genotype of an adapted and already improved cultivar.

Backcrossing is an operation that involves a *recurrent parent* and a *non-recurrent parent*. In many cases the non-recurrent parent is an unadapted line or genotype, and hence is not expected to contribute characters other than the specific one that it is desired to transfer to resulting selections. It is therefore usual to choose a recurrent parent which is already suited to the environment (i.e. the most adapted genotype available). The process of isogenic line production will be identical and repeated for each line that is to be produced.

The process will be described as where the homozygous allele (*RR*) of interest from the non-recurrent parent is completely dominant showing resistance to a disease and the recurrent parent has the recessive completely susceptible alleles (*rr*). First the recurrent parent is crossed to the non-recurrent parent producing F_1 seeds which are therefore heterozygous (*Rr*) for this locus and where each of the two parents contribute equally to the genotype. The F_1s (*Rr*) are crossed back to the recurrent parent *rr* × *Rr*, to produce backcross 1 (BC$_1$). The seeds from this '*backcross*' are of the genotypes *Rr* or *rr*, which can be screened to identify the disease susceptible lines *rr*), as opposed to the disease resistant *Rr*. The *Rr*s can then be used to cross back again to the recurrent parent *rr*, to produce BC$_2$. This process of screening for the presence of the heterozygous resistant lines and backcrossing them to the recurrent parent is repeated a number of times with the aim of developing a line which is comprised of all the genes from the recurrent parent except at the '*resistance locus*' which will have the resistance allele (*R*). In other words we effectively '*add*' this to the genotype. The number of backcrossing generations will depend

on how closely the breeder wants to resemble the recurrent parent or how well the backcross genotypes are performing. The proportion of the recurrent parent genotype in each backcross family will increase with increased backcrossing, and can be calculated by the formula:

$$1 - (1/2)^g$$

where $g =$ the number of backcrossing generations, including the original cross ($P_1 \times P_2$) to produce the F_1. The following are proportions of the genes that are recovered from the recurrent parent according to number of backcrosses:

$$
\begin{aligned}
F_1 &= 50.0\% \\
BC_1 &= 75.0\% \\
BC_2 &= 87.5\% \\
BC_3 &= 93.8\% \\
BC_4 &= 96.9\%
\end{aligned}
$$

The above percentages of the recurrent parent genotype in the resulting progeny hold reasonably well in the early backcross generations. However, with increased backcrossing, the percentage genotype of the non-recurrent parent (the '*wild*' type) will be more influenced by linkage. The resulting backcross inbred can often contain a higher proportion of wild-type genotype than desired, or more backcross operations will be required to obtain the desired proportion of the cultivated/adapted (recurrent) parent.

The backcrossing method where the gene of interest is recessive is slightly more complex (and often a more lengthy) process. The general theory is the same but in this case it is necessary to progeny test the backcrossed generations in order to separate the homozygous and heterozygous plants that need to be selected. Progeny testing can be avoided (or reduced) when tightly linked co-dominant molecular markers are available. Molecular markers also can be used to increase the frequency of the adapted (recombinant) parent genome in the backcross family. This is discussed in more detail in Chapter 8.

DEVELOPMENT OF OUTBREEDING CULTIVARS

Crops that are generally produced as outbred cultivars are: alfalfa, forage legumes, herbage grasses, maizes (some), oil palm, perennial ryegrass, red clover, rye, sugar beet.

Development of open-pollinated crops species is a process that changes the gene frequency of desirable alleles within a population of mixed genotypes while trying to maintain a high degree of heterozygosity. So it is really the properties of the population that is vital not individual genotypes (as in self-pollinating crops). Instead of ending with a cultivar for release that is a uniform genotype the population will be a complex mixture of genotypes, which together give the desired performance.

It is not considered desirable (and often very difficult) to develop homozygous, or near-homozygous breeding lines or cultivars from these outbreeding species as they suffer severe inbreeding depression, carry deleterious recessive alleles or have strong or partial self-incompatibility systems. There are basically two different types of outbreeding cultivars available *open-pollinating populations* and *synthetic cultivars*.

Breeding schemes for open-pollinating population cultivars

In open-pollinating populations, selection of desirable cultivars is usually carried out by *mass selection*, *recurrent phenotypic selection* or *selection with progeny testing*. Outbred (or cross-pollinating) cultivars are maintained through open-pollinated populations resulting from random mating.

Mass selection
Mass selection is a very simple breeding scheme that uses natural environmental conditions to alter the genotypic frequency of an open-pollinating population. A new population is created by cross-pollinating two different existing open-pollinating populations. In this case a representative set of individuals from each population will taken to be crossed – single plants will not of course be representative of the populations. So it is common (even if mistaken) to select individual plants

from each population but to take a reasonable sample of such plants. How they are crossed depends on the choice of the breeder but often Bi-Parental matings are performed (BIPs), where specific parents are selected and hybridized. Alternatively, pollen is often collected in bulk and used to pollinate specifically selected female plants. In many instances breeders allow random mating or open cross pollination to be used.

The seed that results from such a set of crosses is grown under field condition over a number of seasons. The theory of the approach is that genotypes, which are adapted to the conditions, will predominate and be more productive than those that are '*less fit*'. It is also assumed that crossing will be basically at random and result in a population moving towards equilibrium.

Problems with mass selection are related to partial, or complete, lack of control of the environmental conditions other than by choosing suitable locations for the trials. In some instances it is possible to create disease stress by artificially inoculating susceptible plants with the pathogen to act as spreaders, or by growing very susceptible lines in close proximity of the bulk populations. However, the process is empirical and often subject to unexpected disturbances. It also assumes, as noted earlier, that natural selection is going to be in the direction that the breeder desires – an assumption that is not always justified. It should also be noted that care needs to be exercised in isolating this developing population from other crops of this species, which might happen to be growing within pollination distance!

Recurrent phenotypic selection
Recurrent phenotypic selection tends to be more effective than mass selection. The basic outline of this process is illustrated in Figure 4.5. A population is created by cross-pollination between two (or more) populations to create what is referred to as the **base population**. A large number of plants are grown from the base population and a sub-sample of the most desirable phenotypes are identified and harvested as individual plants. These selected plants are then randomly mated to produce a new population – an **improved population**.

This process is repeated a number of times, in other words it is recurrent. The number of cycles performed will be determined by the desired level of improvement required over the base population, the initial gene frequency of the base population and the heritability of

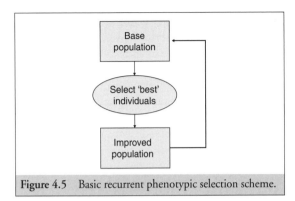

Figure 4.5   Basic recurrent phenotypic selection scheme.

the traits of interest in the selection process. Recurrent phenotypic selection has been shown to be effective but mainly in cases where there is high heritability of the characters being selected for (e.g. some disease and pest resistances). The techniques are not nearly as effective where traits have a lower heritability such as yield or quality traits.

It is common practice (and a good idea) to retain a sample of the base population so that the genetic changes due to selection can be evaluated in a later season.

Progeny testing

There are actually a variety of possibilities within this main heading, for example half- or full-sib selection with test crossing; and selection from S_1 progeny testing. All the schemes basically involve selecting individuals from within the population, crossing or selfing these to produce seed. Part of the seed is sown for assessment and part is retained. Once the results of the assessments are available the remnant seed from the progenies that have been shown to be superior are then sown as a composite population for plants again to be selected and so on. At any stage seeds can be taken out for commercial exploitation.

Backcrossing on open pollinated-cultivar development

Backcrossing is not as commonly used with outbreeding crops as with self-pollinating ones, nevertheless is it used. When used the biggest difference is that the recurrent and non-recurrent parents in the backcrosses are plant populations rather than homozygous lines. The

basic assumption of any backcrossing system is that the technique is unlikely to result in an increase of performance of the recurrent parent, other than for the single character being introduced. Even when a gene has been introduced it is difficult to ensure its distribution over the whole population.

Seed production

In most cases, seed production of open-pollinated cultivars simply involves taking a sample of seed from the population but, to avoid the problems noted before, under increasingly stringent conditions, to avoid contamination or cross-pollination from other populations or cultivars.

DEVELOPING SYNTHETIC CULTIVARS

A synthetic cultivar basically gives rises to the same end result as an open-pollinated cultivar (i.e. population improvement), although a synthetic cultivar cannot be propagated by open-pollination without changing the genetic make-up of the population. This has perhaps been a primary reason for the rapid change over from open-pollinated cultivars to synthetic ones, since it means that farmers need to return to the seed companies for new seed each year. It has been commercial seed companies that have been responsible for breeding almost all synthetic varieties. For example, before 1950 there were only two alfalfa breeders working in private seed companies while 23 were breeding in the public sector. By 1980 there were 17 public sector alfalfa breeders but now there are more than three times (52) private breeders developing synthetic lines.

A synthetic cultivar must be reconstructed from parental lines or clones. Within the United States maize is almost exclusively grown as hybrid cultivars although in many countries maize crops are grown as synthetics. Synthetics have also been used almost exclusively in the development of alfalfa, forage grass and forage legume cultivars, and have also been used to develop varieties from other crop species (e.g. rapeseed).

The breeding method used for the development of synthetic cultivars is dependent on the ability to develop homozygous lines from a species or to propagate parent lines clonally. In the case of maize, for example, synthetic cultivars are developed using a three stage process:

- Develop a number of inbred lines
- Progeny test the inbred lines for general combining ability
- Identify the 'best' parents and intercross these to produce the synthetic cultivar

This process is almost identical to the procedure used for developing hybrid cultivars and only differs in the last stage where many more parents will be included in the synthetic than in a hybrid cultivar. To avoid repetition this section on synthetics will only cover the case of crops where inbred lines are not possible due to inbreeding depression (e.g. clonal synthetics).

The process of developing a synthetic cultivar from clonal lines is illustrated in Figure 4.6.

Clonal selections can be added to the nursery *ad infinitum*, on the basis of continued phenotypic

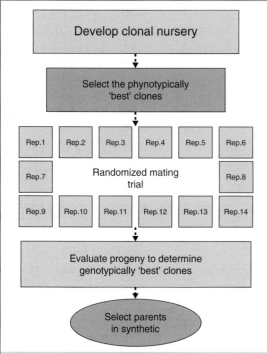

Figure 4.6 Breeding procedure used to develop synthetic cultivars from a clonally reproduced open-pollinated crop species (i.e. alfalfa) and using polycross progeny testing.

recurrent selection on the base population available or by selections from those that have been produced by designed cross pollinations. The second stage involves *clonal evaluation* and is conducted using replicated field trials of asexually reproduced plant units. The aim of the clonal testing is to identify which clonal populations are phenotypically most suited to the environments where they are grown. The clonal trials are often grown in two or more locations to include an assessment of environmental stability.

Using a test cross or polycross technique will then genetically test the '*best*' clones identified from the clonal screen. The aim of this 'genetic' test is to determine the general combining ability of each clonal line in cross combinations with other genotypes in the selected group of clones.

If a test cross (often called a '*top cross*') is used, all the selected clones are hybridized to one (or more) test parent. The test parent will have been chosen because it is a desirable cultivar or it may be chosen because of past experience of the individual breeder. The test parent is a heterozygous clone that produces gametes of diverse genotypes. This diversity of gametes produced from the tester will help an assessment of the average ability of each clone to produce superior progeny when combined with alleles from many different individuals. Test cross evaluations are most useful when the variation that is observed within the different progeny is a result of differences between clones under evaluation and not due to only a small sample of genes coming from the test parent.

A polycross does not use a common test parent but rather a number of different parents. It therefore differs from a test cross as the seed progenies to be evaluated result from inter-crossing between the clones that are under test (i.e. each clone under evaluation is used as female and randomly mated to all, or a good range, of other clonal selections). A polycross, like the test cross, is used to determine the general combining ability of the different clones. The seed so produced from a polycross is then tested in randomized field trials. It is essential that the trials are randomized and that the level of replication is high enough to allow the possibility of hybrid seed being from as many other clones as possible.

Seeds from test crossing and polycrossing are grown in **progeny evaluation trials** to evaluate the genotype or determine the general combining ability of each of the clones. Progeny evaluation trials are very similar to any other plot evaluation trial and are best grown at more than a single environment. Progeny evaluations are also often repeated over a number of seasons to obtain more representative evaluations of the likely performance over different years.

Depending on the results from the progeny evaluation trials, clones that show greatest general combining ability will be used as parents for the synthetic cultivar. These clonal parents are mated in a number of combinations to produce experimental synthetics. The number of clones that are selected and the number of parents that are to constitute the synthetic will determine how many possible combinations are possible. If there were only four clones selected then there would be a total of 11 different synthetic cultivars (that is, the six possible 2-clone combinations, the four 3-clone combinations and the one 4-clone combination). With 6, 8, 10 and 12 parents the number of possible synthetics would be 57, 247, 1013 and 4083, respectively. Therefore it is useful to try to predict the performance of synthetic cultivars without actually producing seed. One formula used is:

$$F_1 - [(F_1 - P)]/n$$

where F_1 is the mean performance of all possible single crosses among n parents, and P is the mean performance of the n parents. It should, however, be noted that there is an assumption of an absence of epistasis (interaction between alleles at different loci) in order to obtain a good estimate of synthetic performance and predictions are therefore often far removed from what is actually observed. Such predictive methods should therefore be used with caution.

Synthetic cultivars have also been developed from out-pollinating species that are semi-tolerant to inbreeding (e.g. maize and rapeseed). In these instances the clonal nursery is replaced by a collection of inbred (or near-inbred) breeding lines, and any other reference (above) to clones is replaced by inbred lines.

Seed production of a synthetic cultivar

The parents used to produce a synthetic cultivar can be inter-crossed by hand-pollination to produce the first generation synthetic (Syn.1). The aim is to cross every parent in the synthetic with all others (i.e. a half diallel cross). This can be difficult with some synthetics (e.g. alfalfa) where the number of parents included is around 40.

It is therefore more common in situations where many parents are used in a synthetic cultivar to produce Syn.1 seed using a polycross procedure. The selected parents are grown in close proximity in randomized block designs with high replication. The crossing block is obviously grown in isolation for any other source of the crop to avoid cross contamination. In cases where insect pollinators are necessary to achieve cross-pollination, attempts are made to ensure that these vectors are available in abundance. For example, alfalfa breeders introduce honey bees or leaf-cutter bees to pollinate synthetic lines.

Seed from Syn.1 is open-pollinated to produce Syn.2, which is subsequently open-pollinated to give the Syn.3 population, etc. The classes of synthetic seed are categorized as breeders' seed, foundation seed and certified seed. In this case breeders' seed would be Syn.1, foundation seed Syn.2 and the earliest certified seed Syn.3.

In summary, the ***characteristics of a synthetic cultivar*** are:

- Synthetic cultivar species need to have potential to have parental lines, which reproduce from source material (either clonally or as an inbred line), and hence the synthetic cultivar can be reproduced from these base parents;
- To develop synthetic cultivars, the contributing parental material is tested for combining ability and/or progeny evaluated;
- Pollination of synthetic cultivar species cannot be controlled, and there has to be some natural method of random mating between parents (e.g. a pollinator or wind pollination);
- The source parental material is maintained for further use;
- Open-pollinated populations have limited life and are then reconstituted on a cyclic basis.

DEVELOPING HYBRID CULTIVARS

Crops that are commonly produced and sold as hybrids are: brussels sprouts, kale, maize, onions, rapeseed, sorghum, sunflower and tomato.

Although attempts have been made to develop hybrid cultivars from almost all annual crop species, it is impossible to consider the evolution of hybrids without a brief history of the developments in hybrid corn. At the beginning of the 20th century it became apparent that genetic advances and yield increases achieved by corn breeders were markedly lower than those realised by other small grain cereal breeders developing wheat and barley cultivars. Indeed the pedigree selection schemes used by corn breeders, although considered suitable at that time, were essentially not effective. The knowledge that hybrid progeny produced by inter-mating two inbred lines often showed heterosis (i.e. produced yields greater than the better parent) suggested that hybrid cultivars could be exploited by corn breeders on an agricultural scale by manually detasseling female parents to produce a population that was entirely comprised of F_1 hybrid seed that could be sold for commercial production.

The first suggestion of using controlled crosses was made by W.J. Beal in the late nineteenth century based specifically on the ideas of Darwin on inbreeding and outbreeding. These ideas were then refined by G.H. Shull in 1909, on the basis of genetic studies, who put forward the idea of a hybrid cross being produced by first developing a series of inbred or near-inbred breeding lines and inter-mating these inbred × inbred crosses (single cross hybrid) and using the hybrid seed for production. These hybrid progeny were indeed high yielding and showed a high degree of crop uniformity. His basic concept, however, was not adopted at that time because the most productive inbred lines had very poor seed yields (most likely due to inbreeding depression) and consequently, hybrid seed production was very expensive.

In the interim the traditional open pollinated corn cultivars were quickly superseded by double cross hybrids ([Parent A × Parent B] × [Parent C × Parent D]) suggested by D.F. Jones in 1918. Double cross hybrids were not as high yielding or as uniform compared to the single cross hybrids proposed by Shull. However, hybrid seed production was less expensive than single cross hybrids and as a result double cross hybrids completely dominated the USA corn production by the 1940's.

Initially all commercial hybrid corn seed was produced by detasseling *female* plants and growing them in close proximity to non-detasseled *males* and only harvesting seed only from the *female* plants. When this method of hybrid corn seed production was most prominent it was estimated that more than 125 000 people were employed in detasseling operations. Increased labour costs combined with developments in using cytoplasmic male sterility (CMS)

production systems in the 1960's, rendered detasseling of females in hybrid corn production obsolete.

From the onset of hybrid corn breeding it was realized that there was a limit to production based on the inbred lines available, and a large effort was put into breeding superior inbred lines to use in hybrid combinations. Introduction of efficient and effective CMS hybrid production systems combined with the development of more productive inbred parents lead to the return of single cross hybrids, which now predominant corn production in the USA, and many other developed countries. The relatively high cost of hybrid corn and other hybrid crop seeds does, however, limit the use of hybrid cultivars in many developing countries.

The rapid increase in popularity and success achieved in hybrid maize could not have occurred without two very important factors. The first is that many countries, including until recently the United States, do (or did) not have any Plant Variety Rights legislation or other means that breeders could use to protect proprietary ownership of the cultivars they bred. There was, therefore, little incentive for private companies to spend time and effort in developing clonal, open-pollinating or inbred cultivars as individual farmers could increase seed stocks themselves, or they could be increased and sold by other seed companies. Hybrid varieties offered the potential for seed/breeding companies to have an in-built economic protection. The developing companies guarded all the stocks of the parental lines and only sold hybrid seed to farmers. These hybrids, although uniform at the F_1 stage, would segregate if seed were retained and re-planted (i.e. they would be F_2 progenies). Secondly, the introduction of hybrid maize occurred simultaneously with the transition from traditional to intensive technology-based agricultural systems. The new hybrids were indeed higher yielding, but were also more adapted to the increased plant populations and rising fertility levels of the times.

There are hardly any agricultural crops where hybrid production has not at least been considered, although hybrids are used in still relatively few crop species. The reasons behind this are first that not all crops show the same degree of heterosis found in maize, and second that it is not feasible in many crop species to find a commercial seed production system that is economically viable in producing commercial hybrid seed. Indeed if maize had not had separate male and female reproductive organs and hence allowed easy female emasculation through detasselling, hybrid cultivar development might never have been developed, or acceptance would have been delayed by at least 20 years, until cytoplasmic male sterile systems were available.

Hybrid cultivars have been developed, however, in sorghum, onions and other vegetables using a cytoplasmic male sterile (CMS) seed production system; in sugar beet and some *Brassica* crops (mainly Brussels sprouts, kale and rapeseed) using CMS and self-incompatibility to produce hybrid seed; and in tomato and potato using hand emasculation and pollination.

If hybrid cultivars are to be developed from a crop, then the species must:

- Show a high degree of heterosis
- Be capable of being manipulated so as to produce inexpensive hybrid seed
- Not easily be produced uniformly, and have a high premium for crop uniformity

There are many differing opinions regarding the exact contribution of hybrids in agriculture. In hybrid maize there would seem little doubt that there have been tremendous advances made. However, this has been the result of much research time and also large financial investments. In addition, it should be noted that inbred parents in hybrid breeding programs have been improved just as dramatically as their hybrid products. Most other hybrid crops (with the exception of sorghum) are also outbreeders. Almost all outbreeding crops show degrees of inbreeding depression and, therefore, its counterpart heterosis. In such cases there are strong arguments, certainly in practical terms, for exploiting heterozygosity to produce productive cultivars. This implies that hybrid cultivars can offer an attractive alternative over open-pollinated cultivars or even synthetic lines, although seed production costs will always be a major consideration. In inbreeding crops, hybrid cultivar production is much more difficult to justify on '*biological grounds*'.

Committed '*hybridists*', of whom there are many (especially within commercial seed companies) would argue that:

- Yield heterosis is there for the exploitation
- Hybrid cultivars are economically attractive to breeding organizations and seed companies

- That technical problems (usually associated with seed production) are simply challenges to be faced
- That the biological arguments are irrelevant

Skeptics (of which there seem to be fewer, or who are less vociferous) argue on the basis of experimental data available to date that:

- There is no good evidence for over-dominance, and so it is definitely possible to develop pure-line, inbred cultivars which are as productive as hybrid lines
- The economic attractions of seed companies should be weighed against high seed costs for the farmers; that technical skills could be put to better (more productive) use
- The biological reality is all-important

These latter skeptics usually, however, accept that in the case of out breeding species hybrids give a faster means of getting yield increases, while in the longer term inbred lines would match them, but in inbreeding crops this differential in speed is not present.

Heterosis

The performance of a hybrid is a function of the genes it receives from both its parents but can be judged by its phenotypic performance in terms of the amount of heterosis it expresses. Many breeders (and geneticists) believe that the magnitude of heterosis is directly related to the degree of genetic diversity between the two parents. In other words, it is assumed that the more the parents are genetically different the greater the heterosis will be. To this end, it is common in most hybrid breeding programmes to maintain two, or more, distinct germplasm sources (**heterotic groups**). Breeding and development is carried out within each source and the different genetic sources are only combined in the actual production of new hybrid cultivars. For example, maize breeders in the United States found that they observed significant heterosis by crossing Iowa Stiff Stalk breeding lines with Lancaster germplasm. Since this discovery, these two different heterotic groups have not been inter-crossed to develop new parental lines but, rather, have been kept genetically separated for parental development.

It has proved difficult to clearly and convincingly define the underlying causes of heterosis in crop plants. There are very few instances where heterozygous advantage *per se* has been shown to result from over-dominance. The counterpart to heterosis, inbreeding depression, is generally attributed to the fixation of unfavorable recessive alleles and so it is argued that heterosis should simply reflect the converse effect. Therefore unfavourable recessive alleles in one line would be masked, in the cross between them, by dominant alleles from the other. If this is all that there is to it, then heterosis should be fixable in true breeding lines. In general it has been found that this simple explanation does not explain all the observed effects. So the question is whether this breakdown in the explanation is related to a statistical problem of the behaviour of a large number of dominant/recessive alleles each with small effect; whether the failure to detect over-dominance is simply a technical failure rather than a lack of biological reality; or whether a more complex explanation needs to be invoked. Dominance can be regarded as the interaction between alleles at the same locus, their interaction giving rise to only one of their products being observed (dominance is expressed) or a mixture of the products of the two (equal mixing giving no dominance, and inequality of mixing giving

different levels of incomplete dominance). But another, well established, type of interaction of alleles can occur, that between alleles at different loci (called non-allelic interaction or epistasis). In addition we cannot simply ignore the fact that linkage between loci is a recognized physical reality of the genetic system that we now regard as being the basis of inheritance. It has been shown that the combination of these two well established genetical phenomena can produce effects that are capable of mimicking the presence of over-dominance.

To examine the effect of having many loci, showing dominance and recessivity, determining the expression of a character differing between the parents of a single cross, let us examine the case of two genetically contrasting parental lines that differ by dominant alleles at only five loci.

Consider the two following cases where we assume that: capital letters represent '*increasing*' alleles, lower case '*decreasing*' ones; each locus contributes in additive fashion to the expression of the character (i.e. the phenotype, let us consider yield); each increasing allele adds the same amount to the yield (2 units) while the decreasing adds nothing to yield; and that dominance is complete and for increasing expression.

Case 1. Parent 1 = AABBCCDDEE and Parent 2 = aabbccddee then:

	Parent 1		**Parent 2**
Genotype	AABBCCDDEE	×	aabbccddee
Phenotype (yield)	$2+2+2+2+2 = 10$		$0+0+0+0+0 = 0$

$$\mathbf{F_1}$$

Genotype	AaBbCcDdEe
Phenotype (yield)	$2+2+2+2+2 = 10$

Case 2. Parent 1 = AAbbCCddEE and Parent 2 = aaBBccDDee then:

	Parent 1		**Parent 2**
Genotype	AAbbCCddEE	×	aaBBccDDee
Phenotype (yield)	$2+0+2+0+2 = 6$		$0+2+0+2+0 = 4$

$$\mathbf{F_1}$$

Genotype	AaBbCcDdEe
Phenotype (yield)	$2+2+2+2+2 = 10$

The level of heterosis is measured by two methods F_1 *minus the mid-parent*, called mid-parent heterosis, or F_1 *minus the best parent*, called best parent heterosis. The mid-parent heterosis need not detain us here as it is of no direct interest. If best parent heterosis is positive (i.e. the F_1 exceeds the performance of the best parent), then generally heterosis has been ascribed to over-dominance.

In Case 1 the best parent (P_1) has a phenotype of 10 yield units; and the F_1 has a phenotype of 10 yield units; that is, they have identical yields and *no heterosis is detected*.

In Case 2 (with the same alleles and effects but with different starting arrangements of alleles between the parents), the best parent (P_1) has a phenotype of 6 yield units; and the F_1 has a phenotype of 10 yield units; that is, *heterosis would be 4 units* and the F_1 is, in fact, 40% more productive than the better parent.

It is however, clearly wrong to consider this to be over-dominance, since none of the individual loci show such an effect. It would certainly be possible (in fact Parent 1 in Case 1 shows this) to fix this level of performance in homozygous, true breeding lines with no heterozygosity (i.e. AABBCCDDEE).

Now let us consider the statistical probability underlying these combinations. With five loci the frequency, assuming no linkage, no effects of selection and a random assortment of gametes in the F_1, after a number of rounds of selfing, the probability (which is equivalent to the frequency) of having a genotype which combines the five dominant alleles as homozygotes would be $(1/2)^5 = 1/32$ (or 0.03125, just over 3%). If, however, you wish to have some assurance that a breeding population contains at least one of these genotype recombinants amongst all possible inbred lines that will be produced, the number of lines needed to be grown and screened is given by the equation:

$$n = \ln(1 - p)/\ln(1 - x)$$

where n is the number of inbred lines that would need to be screened, p is the desired probability that at least one genotype will exist in the population and x is the frequency of the genotype of interest. In the example with five loci, it would be necessary to screen at least 145 inbred lines from the progeny derived from the cross between Parent 1 and Parent 2 (in both cases) to be 99% certain that at least one example of the genotype required will exist.

Of course, a quantitatively inherited trait like yield is not controlled by five loci but more likely 50, 500 or 5000. To give some idea of the number of genotypes needed to obtain a specific combination of alleles when the number of loci increases is given in Table 4.1. In this table the number of plants that need to be screened to be 90%, 95%, 99%, 99.5% and 99.9% sure that at least one genotype of desired combination exists is shown for cases with 5, 6, 7, 8, 9, 10 and 20 loci. With 20 loci, a modest number compared with the possible real situation, a breeder would need to evaluate almost 5 million inbred lines to be sure that the one he wants is present. So hybrids do offer a better probability of success in this instance, but not because they show over-dominance at their loci!

Table 4.1 Number of recombinant inbred lines that would require evaluation for the breeder to be 90%, 95%, 99%, 99.5% and 99.9% sure that a homozygous lines will be a specific combination of alleles at each of 5, 6, 7, 8, 9, 10 and 20 loci.

	Probability of obtaining at least one individual of the desired genotype				
	0.900	0.950	0.990	0.995	0.999
5 Loci	76	94	145	167	218
6 Loci	146	190	292	336	439
7 Loci	294	382	587	676	881
8 Loci	588	765	1 177	1 354	1 765
9 Loci	1 178	1 532	2 356	2 710	3 533
10 Loci	2 357	3 066	4 713	5 423	7 070
20 Loci	2 414 434	3 141 251	4 828 869	5 555 687	7 243 317

A second consideration with hybrid lines that has been postulated is that the heterozygosity of the cultivar makes it more stable over a range of different environments. This may be true but there is no direct evidence for such a basic biological effect and it does not explain the extremely high genotype by environment interactions found in hybrid maize.

Types of hybrid

There are a number of different types of hybrid, apart from the single cross types concentrated on above. The different types of hybrid differ in the number of parents that are in hybrid seed production. Consider four inbred parents (A, B, C and D), types of hybrid would include:

- Single cross hybrids (e.g. A × B, B × D, etc.)
- Three-way hybrids (e.g. [(A × B) × C], [D × (C × A)], etc.)
- Double cross hybrids (e.g. [A × B] × [C × D], [C × A] × [B × D], etc.)

Single cross hybrid are more uniform than three-way or double cross hybrids and are generally more productive, but are most expensive to produce hybrid seed.

Breeding system for hybrid cultivars

The three major steps in producing hybrids are therefore:

- development of inbred lines to be used as parents;
- test cross these lines to identify those that combine well;
- exploit the best single crosses as hybrid cultivars.

The system used to develop hybrid cultivars is illustrated in Figure 4.7. The scheme involves six stages:

- Produce two, or more, segregating populations
- Develop inbred lines (parents)
- Evaluate performance of inbred lines phenotypically
- Evaluate general combining ability of selected inbred lines;
- Evaluate hybrid cross combinations
- Increase inbred parental lines

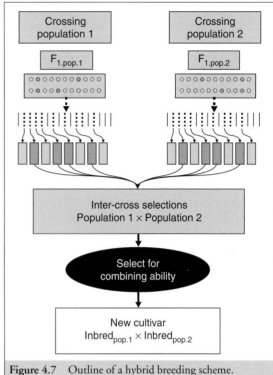

Figure 4.7 Outline of a hybrid breeding scheme.

The procedure used to develop inbred parent lines in hybrid cultivar development is similar to that used to breed pure-line cultivars and the advantages and disadvantages of various approaches are the same. Breeders have used, bulk methods, pedigree methods, bulk/pedigree methods, single seed descent and out of season extra generations (off-station sites) to achieve homozygosity. One of the most important objectives is to maintain high plant vigour and to ensure that the inbred lines are as productive as possible. This is not always easy, particularly in species where there is a high frequency of deleterious recessive alleles present in the outbred populations. Breeders must decide the level of homozygosity that is required. On one hand, the more homozygous (the extreme, of course, being 100% homozygosity) the inbred lines are, then the more uniform will be their resulting hybrid. More heterozygous "inbred lines" may, however, be more productive as parents and hence help reduce the cost of hybrid seed production.

Combining ability (or more relevantly, general combining ability, GCA) is evaluated with the aim of identifying parental lines which will produce productive progeny in a wide range of hybrid cross. Generally, it is not possible to cross all possible parental lines in pair-wise combinations, as the number of crosses to be made and evaluated increases exponentially with the increased number of parents. It is therefore more usual to cross each parent under evaluation to a common **test parent or tester**. The tester used is common to a set of evaluations and therefore, general combining ability is determined by comparing the performance of each progeny, assuming that the only difference between the different progenies can be attributed to the different inbred parents. Testers are usually highly developed inbred lines, which have proved successful in hybrid combinations in the past. A far better prediction of general combining ability would be achieved if more than one tester were used. This, however, is not common practice and breeders have tended to test more inbreds than to increase the number of test parents used.

Evaluation of specific combining ability (or actual individual hybrid combination performance) is carried out when the number of parents is reduced to a reasonable level. The number of possible cross combinations differs with the number of parents to be tested. The number of combinations is calculated from:

$[n(n-1)]/2$ for pair-wise crosses

$[n(n-1)(n-2)]/2$ for three-way crosses

$[n(n-1)(n-2)(n-3)]/8$ for double crosses

where n is the number of parents to be evaluated. For example if 20 parents are to be tested then there would be: 20 crosses to a single tester; 190 pair-wise cross, 3420 three-way cross and 14 535 double cross combinations possible. It is therefore common to predict the performance of three-way and double crosses from single cross performance rather than actually test them. The three-way cross $[(P_1 \times P_2) \times P_3]$ performance is predicted from the equation:

$$1/2[(P_1 \times P_3) + (P_2 \times P_3)]$$

where $P_1 \times P_3$ is the performance of the F_1 progeny from the cross between P_1 and P_3. It is noted that the actual single cross in the hybrid predicted $(P_1 \times P_2)$ is not used in the prediction.

To predict the performance of a double cross $[(P_1 \times P_2) \times (P_3 \times P_4)]$ the following equation is used:

$$1/4[(P_1 \times P_3) + (P_1 \times P_4) + (P_2 \times P_3) + (P_2 \times P_4)]$$

where $P_1 \times P_3$ is the performance of the F_1 progeny from the cross between P_1 and P_3. Note again that the two single crosses used in the double cross do not appear in the prediction equation.

The assumptions underlying this will not be discussed here, but we simply note that this is what is carried out quite often in practical breeding.

Backcrossing in hybrid cultivar development

Backcrossing has featured quite highly in hybrid breeding schemes. Backcrossing is used in hybrid development for two purposes:

- To introduce a single gene into an already desirable inbred parent
- To produce near isogenic lines, which can reduce the cost of seed production, by *convergence*, called *convergent improvement* This is used to make slight improvements to specific hybrid cross combinations and involves recurrent backcrossing between a single cross hybrid ($F_1 = P_1 \times P_2$) and both of its two parents (P_1 and P_2). The result will be to produce from P_1, P_{1*} (where P_{1*}, is a near isogenic line of P_1). Similarly near isogenic lines are developed for P_2. These can be used in a modified single cross ($[P_1 \times P_{1*}] \times P_2$), or a double modified single cross ($[P_1 \times P_{1*}] \times [P_2 \times P_{2*}]$). The aim of this is to increase the efficiency, and reduce cost, of seed production.

Hybrid seed production and cultivar release

The inbred lines used as parents are increased in exactly the same way as pure-line cultivars and hence no further description is needed here.

The first, and highest priority of hybrid seed production is to complete the task as cheaply as possible with the maximum proportion of hybrid offspring. There are four basic means that have been used to produce commercial amounts of hybrid seed:

Mechanical production In hermaphroditic plants, female are emasculated and pollination is achieved either by hand or naturally. In diclinous plants simple physical protection (such as bagging) and hand-pollination are used. The problems with mechanical hybrid production are related to cost, labour inputs and the number of seeds produced by each pollination operation. Examples include tomatoes and potatoes, along with detasselling (as described earlier) in maize.

Nuclear male sterility Nuclear male sterility (NMS) is found in many crop species and sterility is usually determined by the expression of recessive alleles (ms ms) while fertility is present with the dominant homozygote (Ms Ms) and the heterozygotes (Ms ms). Clearly it is not possible to maintain a pure male sterile line by simple sexual reproduction! However, if seeds are collected from male sterile plants they will be heterozygous for the sterility locus and on selfing will segregate 75% fertile (Ms ms or Ms Ms) and 25% sterile (ms ms). The major problems with NMS are related to introducing the necessary genes into commercially desirable genotypes, and the expression being environmentally dependent. Success has been achieved in castor and attempts have been made with cotton, carrot, tomato, sunflower and barley.

Self-incompatibility To produce hybrids one simply inter-plants two self-incompatible, but cross compatible, inbred lines and harvests all the seed, all of which will be hybrid. Problems have been related to overcoming the self-incompatibility of parental lines, in order to maintain them, which can be costly (e.g. bud pollination by hand is often necessary to overcome incompatibility in some *Brassica* species). Incompatibility systems are also often influenced by environmental conditions. Examples of hybrid crops include several vegetables, rapeseed and sunflower.

Cytoplasmic male sterility Cytoplasmic male sterility (CMS) is of wide natural occurrence and can sometimes be induced by backcrossing a desired, adapted genotype (i.e. its nucleus) into a foreign cytoplasm such that the nuclear–cytoplasmic interaction results in male sterility. A practical hybrid system also requires, of course, the restoration of male fertility, usually achieved with 'restorer genes'. These, fortunately, are mostly dominant. If the hybrid product need not produce seeds (e.g. onions or beets) the restorer gene is then not needed. The combination of effective sterility and reliable restorer genes are not often available. Another problem is that the cytoplasm itself (which is always a component of the resulting hybrid) often has deleterious effects on plant viguor. The lack of pollen in the female plants can also have an effect on hybrid production.

For example, if the pollinating vectors are insects they may well prefer to visit only the fertile males. Examples of success are onions, carrot, sugar beet, maize and sorghum.

To produce hybrid cultivars using cytoplasmic male sterility requires three types of genotype: male fertile lines which have a dominant restorer gene (R); cytoplasmic sterile female lines with no restorer gene (called A lines) and "male-fertile" female lines with no restorer genes (called O or B lines). O or B lines are usually isogenic lines of the CMS female parents and are used to maintain and increase seed of the female parent.

DEVELOPMENT OF CLONAL CULTIVARS

Crops that are generally produced as clonal cultivars are: bananas, cassava, citrus, potatoes, rubber trees, soft fruit (raspberry, blackberry, and strawberry), sugarcane, sweet potatoes, top fruit (apples, pears, plums, etc.).

Clonal crops are perennial plant species, although a few clonally propagated crops (e.g. potato and sweet potato) are grown as annual crops. Some clonally propagated crops are very long-lived (e.g. rubber, mango and rosaceous top fruits) and can produce crops for many years after being established. Indeed there are instances whereby fig and palm cultivars have survived over a thousand years, and are still in commercial production. Other clonally propagated crops have a shorter lifespan yet remain in commercial production for several years after being propagated (e.g. sugarcane, bananas, pineapples, strawberries and *Rubus* spp).

There are many methods of propagation used in clonal crop production. Apples, pear, cherry, various citrus, avocado and grape are propagated through budding and grafting onto various root stocks. Leafy cuttings are used to propagate pineapple, sweet potato and strawberry. Leafless stem cuttings are used to propagate sugarcane and lateral shoots are used for banana and palms. There also is, for a number of species, the potential for clonal reproduction via tubers (swollen stems), as is the case for potatoes.

In general, clonal crop species are out-breeders, which are intolerant to inbreeding. Individual clones are highly heterozygous and so it is easy to exploit the presence of any heterosis that is exhibited. Imagine for example that corn could be easily reproduced asexually (i.e. say through apomixis), then there would be no need to develop hybrid corn cultivars because the highly heterozygous nature of a hybrid line could be "genetically fixed" and exploited through asexual reproduction.

The process of developing a clonal cultivar is, in principle, very simple. Breeders generate segregating progenies of seedlings, select the most productive genotypic combination and simply multiply asexually, this also stabilizes the genetic make-up (i.e. avoids problems relating to genetic segregation arising from meiosis). Despite the apparent simplicity of clonal breeding it should be noted that while clonal breeders have shared in some outstanding successes, it has rarely been due to such a simple process, as will be noted from the example below.

Outline of a potato breeding scheme

The breeding scheme that was in use at the Scottish Crop Research Institute prior to 1987 is illustrated in Figure 4.8. It should be noted that the programme used two contrasting growing environments. The **seed site** (indicated by grey boxes) was located at high altitude and was always planted later and harvested earlier than would be considered normal for a typical **ware crop** (indicated by black boxes), in order to minimize problems of insect borne virus disease infection. A ware crop is the crop that is produced for consumption rather than for re-planting.

Each year between 250–300 cross pollinations were carried out between chosen parents. From each cross combination the aim was to produce around 500 seeds, leading to 140 000 seedlings being raised in small pots grown in a greenhouse (two greenhouse seasons were needed to accommodate the 140 000 total). At harvest, the soil from each pot was removed and the tubers produced by each seedling were placed into the now empty pots. At this stage a breeder would visually inspect the small tubers in each pot (each seedling being a unique genotype) and either select or reject each one. One tuber was taken from amongst the tubers produced by the selected seedlings, while all tubers from rejected clones were discarded. The seedling generation, as in most clonal crops, in the greenhouse was the one and only generation that derived directly from *true botanical seeds* (i.e. from sexual reproduction). All

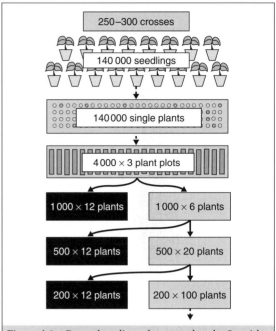

Figure 4.8 Potato breeding scheme used at the Scottish Crop Research Institute prior to 1987.

other generations in the programme were by vegetative (clonal) reproduction, in other words from tubers.

The following year the single selected tubers (approximately 40 000) were planted in the field at the 'seed site' as single plants within progeny blocks. This stage was referred to as the first clonal year. At harvest each plant was harvested, by hand digging, and the tubers exposed on the soil surface in a separate group for each individual plant. A breeder would then visually inspect the produce from each plant and decide, on that basis, to reject or select each group of tubers (i.e. each clone). Three tubers are retained from '*the most desirable*' plants and planted in the field at the same seed site in the following year (the second clonal year) as a three plant, un-replicated plots. First and second clonal year evaluations were therefore carried out under seed site conditions to reduce as far as possible the chances of contamination, especially by virus diseases.

Second clonal year plots were harvested mechanically and, again, tubers from each plot exposed on the soil surface. A breeder examined the tubers produced and decided to select or reject each clone, again on the basis of visual inspection. Tubers from selected clones were

Figure 4.9 Potato yield assessment trials produce a vast volume of product (tubers) from a relatively small area.

retained and grown in the subsequent year for the first time under '*ware growing conditions*' in unreplicated trails at the third clonal year stage. Each selection was also grown as a six plant plot at the seed site. After the second clonal year, however, the seed site was only used to increase clonal tubers and no selection was carried on the basis of the performance at this site.

At the ware site measurements of a variety of characters were taken and yield was recorded (Figure 4.9). Thus the third clonal year, was the first one where selection was based on objective measurements, principally yield but also other performance characters and disease reaction. The fourth and fifth clonal generations were repeats of the third year with reduced numbers of entries after each successive round of selection but with more replicates and larger plots, including with larger multiplication plots of 20 plants and 100 plants, respectively, at the seed site.

In the sixth, seventh and eighth clonal generations surviving clones were evaluated at a number of different locations ('*regional trials*') in the United Kingdom. After each round of trials, the most desirable clones were advanced (i.e. re-trialled) and less attractive clones discarded.

Clones that were selected in each of the three year's regional trials were entered into the UK National List Trials (a statutory government organized national testing scheme). Depending on performance in these trials a decision was made regarding cultivar release and initial foundation seed lots were initiated. If all went well, farmers could be growing newly developed cultivars within 17 years of the initial cross being made.

Time to develop clonal cultivars

Despite the lengthy time period between crossing and farmers growing a new potato cultivar, this is a short time period in comparison to some of the other asexually propagated crops. In potato the long selection process is related to the difficulty in evaluating a crop where the phenotype is greatly affected by the environment (both where the seed and ware crops were grown). In the case of potato, some of the length of the process is related to a slow multiplication rate, around 10 : 1 per generation. In addition, seed tubers are bulky and require large amounts of storage space. To accommodate planting material for one acre of potatoes will require approximately 2000 lbs of seed tubers.

With many other clonal species the time from crossing to cultivar release can be a very lengthy process. In apple breeding, for example, it is often said that if a breeder is successful with the very first parent cross combination, then it is still unlikely that a cultivar will be released (from that cross) by the time the breeder retires! In this case there is the obvious difficulty in the time taken from planting an apple seed to the time that fruit can be evaluated. With several other clonal crops, even the time to develop a new apple cultivar can be considered a relatively short time period in terms of the life of a cultivar. For example, the most common date palm cultivar in commercial production is a clonal line derived from the Middle-east called 'Siguel'. To our knowledge, no one knows the derivation of this line, but the cultivar still predominates as the leading cultivar around the world.

Clonal crop species have shown a high frequency of exploited natural mutations compared to other crop types, probably simply as a result of their clonal propagation making any 'variant' obvious and easily multiplied. As a result many clonal cultivars are the result of natural mutations rather than arising from selection following a specific hybridization between parental lines. For example, the potato cultivar 'Russet Burbank' is a mutation from 'Burbank', and similarly 'Red Pontiac' is a natural mutation of 'Pontiac' and many apple cultivars are simply fruit colour mutants.

Sexual reproduction in clonal crops

In crops that are reproduced from true botanical seed there is definite selection for reproductive normality and high productivity of sexual reproduction. In clonal crops this has not always been the case and the result can hinder the ability of breeders to generate variation by sexual reproduction. There are two main types of clonally reproduced crops (excluding apomicts):

- Those that produce a vegetative product
- Those producing a reproductive product (i.e. fruit)

Those which produce a vegetative product have almost all been selected to have reduced sexual reproductive capacity and can exhibit problems in making sexual crosses. This is probably the result of conscious or subconscious selection for plants which do not 'waste energy' on aspects of sexual reproduction and will therefore put more energy in to the vegetative parts. Extremes are found in yams and sweet potatoes in which many cultivars never flower and in many cases they cannot be stimulated to reproduce sexually. Modern potato cultivars have far less flowering than their wild relatives and of those that do flower, many have very poor pollen viability or are pollen sterile. In the case of potato this

has been the result of conscious human selection, based on the argument noted above and where sexual seed development can also cause a problem of 'volunteers' in subsequent crops.

In clonal crops where the reproductive product is used, there is of course no question about reducing flowering, but nonetheless, reproductive peculiarities and sterility problems are still very common. In general, selection has favoured the vegetative part of fruit development at the expense of seed production. In the extreme, bananas are vegetatively parthenocarpic (i.e. formation of fruit without seeds). Wild bananas are diploid ($n = 11$) and reproduce normally producing fruit with large seeds. Commercially grown bananas are triploid and hence sterile, so many banana cultivars cannot be used as parents in a breeding scheme. Pineapples are also parthenocarpic but self-incompatible, so that clonal plantings give seedless fruits, even though fruit would be seedy if pollinated by another genotype. In addition, mangoes and some citrus sometimes produce polyembryonic plants (where multiple embryos are formed from a zygote by its fission at an early development stage, and hence in effect, results in clonal pseudo-seedlings identical to the mother genotype) which can be a great nuisance to breeders when they are selecting amongst segregating populations of sexually generated progeny.

In conclusion reproductive derangement in clonal species can result in potential sexual crosses between particular parent combinations not being possible, or that individual parental lines cannot be used for sexual reproduction. This limits the options open to breeders of clonally reproduced crops.

Maintaining disease-free parental lines and breeding selections

Several decades ago it was thought that clonal cultivars were subject to **clonal degeneration**, or a reduction in productivity with time. It is now known that this degeneration is primarily the result of clonal stocks becoming infected with bacterial or viral diseases. For example, strawberries can be affected in this detrimental way by infection by bacterial or viral diseases. As an additional example, a bacterial disease causes stunting disease in sugarcane ratoons and not, as was postulated, by genetic '*drift*' causing degeneration. It should, however, be noted that the cause of reduction in performance is often not solely due to disease build up, and some degeneration of clonal lines can be the result of deleterious natural mutations.

Nevertheless, a primary problem in breeding clonal crops is to keep parental lines and breeding stocks free from viral and bacterial diseases that are transmitted through vegetative propagation. Often it is necessary to maintain disease-free breeding stocks that are separated from those that are being tested for adaptation potential, as was the case with the potato breeding scheme described above.

Seed increase of clonal cultivars

One of the primary concerns associated with the increase of new clonal cultivars is to prevent the stocks becoming infected by viral or bacterial diseases during commercial increase. *In vitro* methods can often help in providing rapid increases of clonal stocks, which can be easily kept disease-free. Such methods cannot always be easily applied to many (particularly woody plant types) species, where rates of multiplication *in vitro* are often low and re-establishment under normal field conditions is not always possible. *In vitro* multiplication and applications for disease-free clonal material are discussed later.

DEVELOPING APOMICTIC CULTIVARS

It is possible to develop clonal cultivars through obligate apomixis (e.g. as with buffel grass). In species that are grown from obligate apomictic seeds there tends to be a positive relationship between the productivity of a clone and its level of heterozygosity. It is therefore desirable to develop sources of genetic variability to maximize productivity within the population and this is a continual challenge to breeders of apomictic crops. Essentially all the seed produced by an obligate apomict are produced as a result of asexual reproduction. By far the most common means of producing genetic variation and/or increasing heterozygosity, however, is by sexual reproduction or hybridization between two chosen lines or populations. In developing apomictic cultivars it is important that some seed can be sexually produced so that this type of variation can be exploited.

There are basically two methods of producing apomictic cultivars.

- Selfing a sexual clone to produce a segregating S_1 (due to the clone being highly heterozygous). The progeny can be screened for sexuals and apomicts. If apomicts are obligate they have the potential of being a clonal cultivar.
- Crossing a sexual clone to an apomict clone to produce an 'F_1' and selecting for sexual or apomicts within the segregating population. Obligate apomicts have the potential of being new cultivars.

The method is illustrated in Figure 4.10. Of course a breeding scheme can use a combination of both methods of selfing and inter-crossing to develop a breeding system. In many cases lines have been developed by first selfing clones to achieve some degree of inbreeding, which is aimed at reducing the frequency of deleterious alleles, and then inter-crossing the partial inbred lines to develop the apomict cultivar. The scheme shown in Figure 4.10 also could begin by selfing a semi-apomictic selection rather than a cross between an apomict and a sexual parent.

Once the apomicts have been identified, the most adapted lines are selected through repeated rounds of evaluation over different environments and years.

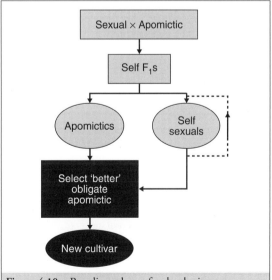

Figure 4.10 Breeding scheme for developing near obligate apomict cultivars.

SUMMARY

In summary, different crop species are more, or less, suited for developing different cultivar types. Individual plant breeders have developed specific breeding schemes that best suit their situation according to the type of cultivar being developed, the crop species and the resources (financial and others) that are available.

Irrespective of which exact scheme is used, breeding programmes also differ in the number of individual phenotypes that are evaluated in each scheme and in the characters that are used for selection at each of the breeding stages. To address the issue of numbers in the breeding program, it is necessary to consider the mode of inheritance of the factors of interest, and to have an understanding of the relationship between the genetics of their inheritance.

THINK QUESTIONS

(1) Disease resistance to powdery mildew has been identified in a primitive and uncultivated relative of barley (*Hordeum vulgare*). Research has shown that the resistance is controlled by a single dominant allele (notation = *RR*). There is a 100% reliable resistance screen test available. Diagrammatically, outline a breeding scheme that could be used to introgress this character into an existing cultivar ('Golden Promise') which is susceptible to powdery mildew (notation = *rr*). The aim is that the resulting resistant lines are to have at least 96% of the Golden Promise genotype.

Consider now that the disease resistance is controlled by a single recessive allele, how would this effect the breeding method you have described above.

(2) Crop cultivars can be divided into pure-line cultivars, cross-pollinated cultivars, hybrid cultivars or clonal cultivars. Briefly, describe the major problems that can be encountered or the attributes of the crop types that can be utilized, when breeding: a pure-line cultivar; a cross-pollinated cultivar; a hybrid cultivar and a clonal cultivar.

(3) You have been appointed as alfalfa (*Medicago sativa* L.) breeder to develop crops for the Alfred Alfoncia Seed Company Ltd. Outline a suitable breeding scheme you will use to develop superior synthetic cultivars for your new employer. Indicate each stage of the scheme on a year-by-year basis.

After several rounds of clonal selection, you have identified 4 potential high production parent lines for developing a new synthetic (lines are coded as **A**, **B**, **C** and **D**). F_1 hybrid seed was produced from all parent cross combinations possible and these F_1s were evaluated in yield trials along with the parent lines. From the yield (Kg) results (below) indicate the **three parent** synthetic most likely to be highest yielding and state the expected yield.

Parent A = 18; Parent B = 22;

Parent C = 21; Parent D = 25

A × B = 25   A × C = 26   A × D = 24

B × C = 30   B × D = 21   C × D = 24

(4) Outline three characteristics of a crop species that would merit consideration for developing hybrid cultivars from that species.

(5) Outline the advantages and disadvantages of a bulk breeding scheme and a pedigree breeding scheme to develop inbred line cultivars

Many breeding schemes for inbred lines are a combination of bulk and pedigree schemes. Design a suitable bulk/pedigree breeding scheme (it is not necessary to make notes on when specific characters are selected) that could be used to develop superior inbred line cultivars. Outline any advantages or disadvantages of your breeding scheme.

(6) A hybrid breeding programme has identified six superior inbred lines (A, B, C, D, E and F) and these were inter-crossed in all combinations and the F_1 populations evaluated for productivity in a field yield trial. The results from each single cross in the trial were:

A × B = 34   A × C = 29   A × D = 27

A × E = 33   A × F = 20   B × C = 31

B × D = 34   B × E = 26   B × F = 19

C × D = 29   C × E = 30   C × F = 18

D × E = 35   D × F = 10   E × F = 19

Which parent combination will provide the most productive **single cross** hybrid and what would the expected yield be? Which parent combination will provide the most productive **three-way cross** hybrid and what would the expected yield be? Which parent combination will provide the most productive **double cross** hybrid and what would the expected yield be?

Explain why the most productive double cross combination (from the prediction equation), may not result in the most commercially suited hybrid cultivar.

(7) In order to justify development of hybrid cultivars it is necessary to have some method of producing inexpensive hybrid seed. Outline three systems that have been used to produce hybrid seed and indicate the advantages or disadvantages of each system.

(8) 'Russet Burbank' has been the predominant potato cultivar in the Pacific Northwest for many years. You have been asked to design a breeding programme that will produce cultivars that can replace some of the Russet Burbank acreage. Outline the breeding design you would use, indicating: (a) five breeding objectives; the breeding scheme (use diagrams if necessary), indicating number of clones evaluated and selected at each stage.

(9) In order to produce a hybrid cultivar using cytoplasmic male sterility you require three types of genotypes. Describe the difference between the three types required, and describe how commercial production will be conducted.

(10) You are to embark on a backcrossing study designed towards convergence breeding. You have at your disposal two genotypes coded C and M. Genotype C is a cytoplasmic male sterile line, but one that does not have very good hybrid combining ability; genotype M has great hybrid combining ability and is male fertile (without a male fertile restorer gene). Design a scheme (listing each operation necessary) that will result in a genotype which is 95% the genotype of the M line but is cytoplasmically male sterile.

(11) Mass selection and recurrent phenotypic selection are often used in developing out-pollinating cultivars. Describe the main difference between the selection methods and indicate any problems that each selection method could have in a breeding programme.

(12) You are walking along a deserted beach in Southern California (are there any of these left?) and you inadvertently kick a brass oil lamp. On picking up the lamp you gently rub the side and – *poof!* – a 20 foot tall genie appears! '*Oh master!*' says the genie, '*I can grant you two wishes*' (well times are hard and the old three wishes thing does not apply any more). You think for a few minutes then say '*I would like you to conjure me up the most wonderful plant that exists within the Universe, so that I can grow it on my farm in Idaho and make millions of dollars in profits*'. *Poof!* This plant appears before you. A plant species that has never appeared on earth before.

As a plant breeder, list five questions you would ask the genie about the biology of this plant that would help you design a cultivar development programme to increase the value of it as a crop.

Having sorted that out, you begin to think of your other wish. '*I would like you to tell me the formula for chemical apomicts*' (a chemical, when applied to a crop will result in 100% apomictic seed from the crop). '*Well*' says the genie. '*I can do this, but, you must specify which crop species the chemical will work on*' (it can only work on one chosen species). What crop species would you chose to have apomictic seeds, and why?

5

Genetics and Plant Breeding

INTRODUCTION

Early plant breeders, basically farmers, did not have any knowledge of the inheritance of characters in which they were interested. The only knowledge they possessed was that the most productive offspring tended to originate from the most productive plants and that the better flavour types tended to be derived from parent plants which were, themselves, of better flavour. Nonetheless, the achievements of these breeders were considerable and should never be underestimated. They moulded most of the crops as we recognize them today from their wild and weedy ancestral types.

In modern plant breeding schemes it is recognized, however, that it is very much more effective and efficient (or indeed essential) to have a basic knowledge of the inheritance or genetics of traits for which selection is to be carried out.

There are generally five different areas of genetics that have been applied to plant breeding:

- **Qualitative genetics**, where inheritance is controlled by alleles at a single locus, or at very few loci
- **Population genetics**, which deals with the behaviour (or frequency) of alleles in populations and the conditions under which they remain in equilibrium or change. Thus allowing predictions to be made about the properties and changes expected in populations
- **Quantitative genetics**, for traits where the variation is determined by alleles at more than a few loci, traits that are said to be controlled by polygenic systems. Quantitative genetics is concerned to describe the variation present in terms of statistical parameters such as progeny means, variances and covariances
- **Cytogenetics**, the study of the behaviour and properties of chromosomes being the structural units which carry the genes that govern expression of all the traits

- **Molecular genetics**, where studies are carried out at the molecular level. Molecular techniques have been developed to investigate and handle both qualitative and quantitative characters. Although the details of molecular genetics are generally outside the scope of this book, the impact of molecular genetic techniques is important and relevant

QUALITATIVE GENETICS

Few sciences have as clear-cut a beginning as modern genetics. As mentioned previously, early plant breeders were aware of some associations between parent plant and offspring and at various times in history researchers had carried out experiments to study such associations. However, experimental genetics with real meaning began in the middle of the 19th century with the work of Gregor Johann Mendel, and only fully appreciated after the turn of the 20th century.

Mendel's definitive experiments were carried out in a monastery garden on pea lines. The flowers of pea plants are so constructed as to favour self-pollination and as a result the majority of lines used by Mendel were either homozygous or near-homozygous genotypes. Mendel's choice of peas as an experimental plant species offered a tremendous advantage over many other plant species he might have chosen. The differences in characters he chose were also fortuitous. Therefore many present day scientists have argued that Mendel had a great deal of luck associated with his findings because of the choices he made over what to study. When this is combined with segregation ratios that are better than might be expected by chance, many have concluded that he must 'have already foreseen the results he expected to obtain'.

Before considering an example of one of Mendel's experiments, there are a few general points to be made of his experiments. Others, before Mendel, had made controlled hybridizations or crosses within various species. So why did Mendel's crosses, rather than those of earlier workers, provide the basis for the modern science of genetics? First and foremost Mendel had a brilliant analytical mind that enabled him to interpret his results in ways that defined the principles of heredity. Second, Mendel was a proficient experimentalist. Therefore he knew how to carry out experiments in such a way as to maximize the chances of obtaining meaningful results. He knew how to simplify data in a meaningful way. As parents in his crosses, he chose individuals that differed by sharply contrasting characters (now known to be controlled by single genes). Finally, as noted above, he used true breeding (homozygous) lines as parents in the crosses he studied.

Among other elements in Mendel's success were the simple, logistical sequence of crosses that he made and the careful numerical counts of his progeny that he kept with reference to the easily definable characteristics on whose inheritance he focussed his attention. It should be noted that many of the features listed as reasons for Mendel's success are very similar to the criteria necessary to carry out a successful plant breeding programme (including the '*luck element*').

Let us consider some of Mendel's experiments as an introduction to qualitative inheritance. These results are presented in Table 5.1. When Mendel crossed plants from a round-seeded line with plants from a wrinkled-seeded line, all of the first generation (F_1) had round seeds. The characteristic of only one of the parental types was therefore represented in the progeny. In the next generation (F_2), achieved by selfing the F_1, both round-seed and wrinkled-seed were found in the progeny. Mendel's count of the two types in the F_2 was

5474 round-seed and 1850 wrinkled-seed, a ratio of $2.96 : 1$.

Mendel found it notable that the same general result occurred when he made crosses between plants from lines differing for other characters. Another example is when he crossed peas with yellow cotyledons with ones with green cotyledons the F_1s all had yellow cotyledons while he found a ratio of about 3 yellows to 1 green in the F_2. Almost identical results were obtained when long-stemmed plants were crossed with short-stemmed plants and when plants with axial inflorescence were crossed to plants with terminal inflorescence.

What did Mendel make of his generalized findings? One of the keys to his solution was his recognition that in F_1 the heredity basis for the character that fails to be expressed is not lost. This expression of the character appears again in the F_2 generation. Recognizing the idea of dominance and recessiveness in heterozygous genotypes and the particulate nature of the heritable factors was the overwhelming genius of Gregor Mendel. This laid the foundation of genetics and hence the explanation underlying the most important features of qualitative genetics.

Genotype/phenotype relationships

Within genetic studies there are two inter-related points. The first is concerned with the actual genetic make-up of individual plants or segregating populations and is referred to as the genotype. The second is related to what is actually expressed or observed in individual plants or segregating populations and is termed the phenotype.

In the absence of any environmental variation (which can often be assumed with qualitative, single, major gene, inheritance) the most frequent cause of difference between genotype and phenotype is due to **dominance**

Table 5.1 Results from some of Mendel's crossing experiments with peas.

Phenotype of parents	F_1 progeny	# of F_2 progeny	F_2 ratio, dominant : recessive by phenotype
round (r) × wrinkled (w), seed	round	5,474 e : 1850 w	2.96 : 1
Yellow (y) × green (g), coty.	yellow	6,022 y : 2001 g	3.01 : 1
long (lo) × short (sh), stem	long	787 lo : 277 sh	2.84 : 1
axil (ax) × terminal (ter), inf	axil	651 ax : 207 ter	3.15 : 1

effects. For example, consider a single locus and two alleles (A and a). If two diploid homozygous lines are crossed where one has the genotype AA and the other aa, then the F_1 will be heterozygous at this locus (i.e. Aa). If this resembles exactly the AA parent then allele A is said to be dominant over allele a. On selfing the F_1, genotypes will occur in the ratio $\frac{1}{4}AA : \frac{1}{2} Aa : \frac{1}{4} aa$. If however, the allele A is completely dominant to a, the phenotype of the F_2 will be $\frac{3}{4}A : \frac{1}{4}a$. Therefore **complete dominance** occurs when the F_1 shows exactly the same phenotype as one of the two parents in the cross.

Compare these ratios to those observed by Mendel (Table 5.1). You will notice that Mendel's data do not fit exactly to a $3 : 1$ ratio that would be expected, given that each trait is controlled by a single completely dominant gene. It should, however, be remembered that gamete segregation in meiosis and pairing in fertilization are **random events**. Therefore it is highly unlikely, because of sampling variation that exact ratios are found in such experiments. Indeed many more recent genetic researchers have found that Mendel's data appears to be 'too good a fit' for purely random events. Whether these geneticists are correct or not does, however, not detract in any way from the remarkable analytical mind of Mendel.

Segregation of qualitative genes in diploid species

To illustrate segregation of qualitative genes, and indeed many following concepts, consider a series of simple examples. First, consider two pairs of single genes in spring barley (*H. vulgare* L.). Say that dwarf barley plants differ in plant height from tall types at the t-locus with tall types given the genetic constitution TT, and dwarf lines tt. 6-row barley ears differ from 2-row ears, where the central florets do not set, at the S-locus, with 6-row SS being dominant over 2-row types, ss. Let us consider the case where two homozygous lines are artificially hybridized where one parent is homozygous tall and 6-row and the other is dwarf and 2-row. First consider the segregation expected of each trait.

Parents	Tall × Dwarf	6-row × 2-row
F_1	Tall	6-row

When the tall F_1 is backcrossed to the dwarf parent, a ratio of 1 tall : 1 dwarf is obtained in the resulting progeny. Similarly, when the six-row F_1 is crossed to a homozygous two-row a ratio of 1 six-row : 1 two-row is obtained. Evidently, the difference between tall and dwarf behaves as a single major gene with the tall allele being completely dominant to the dwarf allele; and the difference between six-row and two-row is also a single gene with six-row being completely dominant to two-row. In terms of segregating alleles (i.e. genotypes), the above example would be:

Parents	$TT \times tt$	$SS \times ss$
F_1	Tt	Ss
F_2	$TT : Tt : tt$	$SS : Ss : ss$
	$1 : 2 : 1$	$1 : 2 : 1$

Where TT and Tt (or SS and Sc) have the same phenotype, it results in the $3 : 1$ phenotypic segregation ratio. If each allele exerts equal effect (additive, and no dominance) then we would have expected three phenotypes in the ratio of 1 tall, 2 intermediate and 1 short, according to the $1\ TT : 2Tt : 1\ tt$ Mendelian ratio.

Now assume that a large number of the F_2 plants were allowed to set selfed seed, what would be the ratio of phenotypes and genotypes in the F_3 generation? To answer this consider that only the heterozygous F_2 plants will segregate (i.e. TT and tt types are now homozygous for that trait and will breed true for that character). Of course the heterozygous F_2 has the same genotype as the F_1 and so it is no surprise that it segregates in the same ratio. At F_3 we therefore have $(1/4TT + 1/2 \times 1/4TT)$: $1/2 \times 1/2Tt$: $(1/2 \times 1/4tt + 1/4tt)$ which results in $3/8\ TT : 2/8\ Tt : 3/8\ tt$. Similar results of $3/8\ SS : 2/8\ Ss : 3/8\ ss$ would be obtained for 6-row versus 2-row. Expanding this to later generations we have:

F_4	$7/16\ TT$	:	$2/16\ Tt$	:	$7/16\ tt$
F_5	$15/32\ TT$	:	$2/32\ Tt$	:	$15/32\ tt$
F_6	$31/64\ TT$	:	$2/64\ Tt$	:	$31/64\ tt$
$\vdots$					
F_∞	$1/2\ TT$	:	$0\ Tt$	:	$1/2\ tt$

You note that each successive generation of selfing reduces the proportion of heterozygotes by half (i.e. 1/2 *Tt* at F$_2$, 2/8 *Tt* at F$_3$, 2/16 *Tt* at F$_4$ and 2/32 *Tt* at F$_5$).

Let us now consider how these two different pairs of alleles behave with respect to each other in inheritance. One way to study this is to cross individuals that differ in both characteristics:

tall, 6-row × dwarf, 2-row

(*TTSS* × *ttss*)

When this cross is made, the F$_1$ shows both dominant characteristics, tall and 6-row. Plant breeder's interests in genetics mainly relate to selection and as no selection takes place at the F$_1$ stage so the interest begins when the self-pollinated progeny of the F$_1$ is considered. F$_1$ individuals are assumed to produce equal frequencies of four kinds of gametes during meiosis (*TS, Ts, tS* and *ts*). An easy way to illustrate the possible combination of F$_2$ progeny is using a **Punnett square**, where the four gamete types from the male parent are listed in a row along the top, and the four kinds from the female parent are listed in a column down the left hand side. The 16 possible genotype combinations are then obtained by filling in the square, that is:

Gametes from female parent	Gametes from male parent			
	TS	*Ts*	*tS*	*ts*
TS	*TSTS*	*TSTs*	*TStS*	*TSts*
Ts	*TsTS*	*TsTs*	*TStS*	*Tsts*
tS	*tSTS*	*tSTs*	*tStS*	*tSts*
ts	*tsTS*	*tsTs*	*tstS*	*tsts*

This also can be written as:

Gametes from female parent	Gametes from male parent			
	TS	*Ts*	*tS*	*ts*
TS	*TTSS*	*TTSs*	*TtSS*	*TtSs*
Ts	*TTSs*	*TTss*	*TtSs*	*Ttss*
tS	*TtSS*	*TtSs*	*ttSS*	*ttSs*
ts	*TtSs*	*Ttss*	*ttSs*	*ttss*

Collecting the like genotypes we get the following frequency of genotypes:

1/16 *TTSS*; 2/16 *TTSs*; 1/16 *TTss*

2/16 *TtSS*; 4/16 *TtSs*; 2/16 *Ttss*

1/16 *ttSS*; 2/16 *ttSs*; 1/16 *ttss*

and with the frequency of phenotypes:

9/16 *T_S_* : = tall and 6-row

3/16 *T_ss* : = tall and 2-row

3/16 *ttS_* : = short and 6-row

1/16 *ttss* = short and 2-row

As with the single gene case above, if the alleles do not show dominance then there would indeed be nine different phenotypes in the ratio:

1 *TTSS* : 2 *TTSs* : 1 *TTss*

2 *TtSS* : 4 *TtSs* : 2 *Ttss*

1 *ttSS* : 2 *ttSs* : 1 *ttss*

In most cases when developing inbred cultivars, plant breeders carry out selection based on plant phenotype amongst early generation (F$_2$, F$_3$, F$_4$, etc.), segregating populations. It is obvious that 75% of the F$_2$ plants will be heterozygous at one, or both, loci, and dominance effects can mask the true genotypes that are to be selected. Single F$_2$ plant selections for the recessive traits (short stature and 2-row) allows for identification of the desired genotype (*ttss*) but only 1/16th of F$_2$ plants will be of this type, while the recessive expression of the trait in most plants is not expressed due to dominance.

The effects of heterozygosity on selection can be reduced through successive rounds of self-pollination. Plant breeders, therefore, do not only select for single gene characters at the F$_2$ stage. Consider now what would happen if a sample of F$_2$ plants from the above example were selfed, what then would be the resulting segregation expected at the F$_3$ stage?

There are nine different genotypes at the F$_2$ stage, *TTSS, TTSs, TTss, TtSS, TtSs, Ttss, ttSS, ttSs* and *ttss* and that they occur in the ratio 1 : 2 : 1 : 2 : 4 : 2 : 1 : 2 : 1, respectively. Obviously if any genotype is homozygous at a locus then these plants will not segregate at that locus. For example plants with the genetic constitution of *TTSS* will always produce plants with the *TTSS* genotype. F$_2$ plants with a genotype of *TTSs* will not segregate for the *Tt* locus and will only segregate for the *Ss* locus. Similarly, F$_2$ plants with a genotype of *TtSs* will segregate for both loci, with the same segregation frequencies as the F$_1$.

From this we have:

F$_3$	F$_2$									
	TTSS	*TTSs*	*TTss*	*TtSS*	*TtSs*	*Ttss*	*ttSS*	*ttSs*	*ttss*	Total
	4/64	8/64	4/64	8/64	16/64	8/64	4/64	8/64	4/64	
TTss	4	2	–	2	1	–	–	–	–	9
TTSs	–	4	–	–	2	–	–	–	–	6
TTss	–	2	4	–	1	2	–	–	–	9
TtSS	–	–	–	4	2	–	–	–	–	6
TtSs	–	–	–	–	4	–	–	–	–	4
Ttss	–	–	–	–	2	4	–	–	–	6
ttSS	–	–	–	2	1	–	4	2	–	9
ttSs	–	–	–	–	2	–	–	4	–	6
ttss	–	–	–	–	1	2	–	2	4	9
										64

Summing over rows, this results in a genotypic segregation ratio of 9/64 *TTSS* : 6/64 *TTSs* : 9/64 *TTss* : 6/64 *TtSS* : 4/64 *TtSs* : 6/64 *Ttss* : 9/64 *ttSS* : 6/64 *ttSs* : 9/64 *ttss*.

The phenotypic expectation at the F$_3$ would be:

$$T\_S\_ = 25/64 \ = \text{Tall and 6-row}$$

$$T\_ss = 15/64 \ = \text{Tall and 2-row}$$

$$ttS\_ = 15/64 \ = \text{short and 6-row}$$

$$ttss = 9/64 \ \ = \text{short and 2-row}$$

This result could be obtained in a simpler manner by considering the segregation ratio of each trait separately and using these to form a Punnett square. For example, the frequency of homozygous tall (*TT*), heterozygous tall (*Tt*) and homozygous short (*tt*) at the F$_2$ is 1 : 2 : 1, respectively. Similarly, the frequency of *SS*, *Ss* and *ss* is also 1 : 2 : 1. We can use these frequencies to construct a Punnett square such as:

Gametes	*TT* – 1/4	*Tt* – 1/2	*tt* – 1/4
SS – 1/4	1/16 *TTCC*	2/16 *TtCC*	1/16 *ttCC*
Ss – 1/2	2/16 *TTCc*	4/16 *TtCc*	2/16 *ttCc*
ss – 1/4	1/16 *TTcc*	2/16 *Ttcc*	1/16 *ttcc*

This would result in the same genotypic and phenotypic frequencies as shown above. It is, however, necessary to be familiar with the more direct method for situations where segregation is not independent (i.e. in cases of linkage).

It should be noted, as in the single gene case, increased selfing results in increased homozygosity. Therefore with increased generations of selfing there is an increase in the frequency of expression of recessive traits. If, in this case, a breeder wished to retain the 6-row short plant type, then it would be expected that 3/16 (approximately 19%) of all F$_2$ plants will be of this type (*ttS_*), and only 1/3 of these would at this stage be homozygous for both traits. If selection were delayed until the F$_3$ generation, then 15/64 (or just over 23%) of the population would be of the desired type, and now 60% of the selections would be homozygous for both genes. Obviously, if selection is delayed until after an infinite amount of selfing, then the frequency of the desired types would be 25% (and all homozygous).

Qualitative linkage

The principle of independent assortment of alleles at different loci is one of the corner stones on which an understanding of qualitative genetics is based. Independent assortment of alleles does not, however, always occur. When certain different allelic pairs are involved in crosses, deviation from independent assortment regularly occurs if the loci are located on the same chromosome. This will mean that there will be a tendency of parental combinations to remain together, which is expressed in the relative frequency of new combinations, and is the phenomenon of *linkage*.

It is often desirable to have knowledge of linkage in plant breeding to help predict segregation patterns in various generations and to help in selection decisions. All genetic linkages can be broken by successive rounds of sexual reproduction, although it may take many attempts of recombination before tight linkages are broken. In general, however, if two traits of interest are adversely linked (i.e. they appear jointly with lower than expected frequency) then increased opportunities for recombination need to take place before selection.

The ratio of F$_2$ progeny after selfing F$_1$ individuals (i.e. say *AaBb*), of equal numbers of gametes of the four possible genotypic combinations (*AB*, *Ab*, *aB*, *ab*), leads to the ratio of $9AB : 3Ab : 3aB : 1ab$. Similarly, we find the genetic ratio of $1AB : 1Ab : 1aB : 1ab$, on test crossing the F_1 to a complete recessive (aabb). Any deviation from these expected ratios is an indication of linkage between the two gene loci.

Consider a second simple example of recombination frequencies derived from a test cross, and how the test cross can be used to estimate recombination ratios and hence ascertain linkage between characters. Consider the dwarfing gene in spring barley and a third single gene, which confers resistance to barley powdery mildew. Mildew resistant genotypes are designated as RR and susceptible genotypes as rr. A cross is made between two barley genotypes where one parent is homozygous tall and mildew resistant (i.e. $TTRR$); and the other is short and mildew susceptible (i.e. $ttrr$). We would expect the F_1 to be tall and resistant (i.e. $TtRr$). When the heterozygous F_1 was crossed to a completely recessive genotype (i.e. $ttrr$) we would expect to have a $1 : 1 : 1 : 1$ ratio of phenotypes. When this cross was carried out and the progeny from the test cross was examined the following frequencies of phenotypes and genotypes were the ones actually observed.

Tall and Resistant ($TtRr$)	= 79
Short and Resistant ($ttRr$)	= 18
Tall and Susceptible ($Ttrr$)	= 22
Short and Susceptible ($ttrr$)	= 81
Total	**= 200**

Linkage is obviously indicated by these results as the observed frequencies are markedly different from those expected at $50 : 50 : 50 : 50$, on the basis of independent assortment. The two phenotypes, *tall and resistant* and *short and susceptible*, occur at a considerably higher frequency than we expected. You will readily note that these are the same phenotypes as the two parents. The other two phenotypes (the non-parental types) were observed with much lower frequency. Added together, the two recombination types (*short and resistant*, and *tall and susceptible*) only account for 40 (20%) out of the 200 F_2 progeny examined, while we expected their contribution to collectively make up 50% of the progeny. Therefore the F_1 plants are producing TR gametes or tr gametes in 80% of meiotic events, or a frequency of 0.4 for each type. Similarly Tr or tR gametes are produced

in only 20% of meiotic events, or a frequency of 0.1 for each type. Knowing the frequency of four gamete types we can now proceed and predict the frequency of genotypes and phenotypes we would expect in the F_2 generation using a punnet square.

Gametes from female parent	Gametes from male parent			
	$TR - 0.1$	$Tr - 0.4$	$tR - 0.4$	$tr - 0.1$
$TR - 0.1$	$TTRR$ 0.01	$TTRr$ 0.04	$TtRR$ 0.04	$TtRr$ 0.01
$Tr - 0.4$	$TTRr$ 0.04	$TTrr$ 0.16	$TtRr$ 0.16	$Ttrr$ 0.04
$tR - 0.4$	$TtRR$ 0.04	$TtRr$ 0.16	$ttRR$ 0.16	$ttRr$ 0.04
$tr - 0.1$	$TtRr$ 0.01	$Ttrr$ 0.04	$ttRr$ 0.04	$ttrr$ 0.01

Collecting like genotypes results in:

Tall and resistant	$= T\_R\_$	$= 0.66$
Short and resistant	$= ttR\_$	$= 0.09$
Tall and susceptible	$= T\_rr$	$= 0.09$
Short and susceptible	$= ttrr$	$= 0.16$

From a breeding standpoint compare these phenotypic ratios to those that would have been expected with no linkage (i.e. $0.56T\_R\_ : 0.19ttR\_ : 0.19 T\_rr : 0.06ttrr$). If, as might be expected, the aim was to identify individual plants which were short and resistant, then the occurrence of these types would drop from 19% to 9%, by more than half. It should also be noted that 10% recombination is not particularly high.

Overall, therefore, genes on the same chromosome (particularly same chromosome arm) are linked and do not segregate independently. Two heterozygous loci may be linked in coupling or repulsion. Coupling is present when desirable alleles at two loci are present together on the same chromosome and the unfavourable alleles are on another (e.g. TR/tr). Repulsion is present when a desirable allele at one locus is on the same chromosome as an unfavourable allele at another locus (e.g. Tr/tR).

In the test cross $TtRr \times ttrr$ the situations shown in Table 5.2 can arise.

The percentage recombination, expressed as the number of map units between loci, can be calculated

Table 5.2 Expected frequency of genotypes resulting from a test cross where *TrRr* is crossed to *ttrr* with independent assortment, and with coupling and repulsion linkage.

Independent assortment	Coupling linkage	Repulsion linkage
(1/4) *TR*	>(1/4) *TR*	<(1/4) *TR*
(1/4) *Tr*	<(1/4) *Tr*	>(1/4) *Tr*
(1/4) *tR*	<(1/4) *tR*	>(1/4) *tR*
(1/4) *tr*	>(1/4) *tr*	<(1/4) *tr*

from the equation:

Recombination %

$$= \frac{\text{sum of non-parental classes}}{\text{total number of individuals}} \times \frac{100}{1}$$

This depending on what the initial parents were will be estimated as:

$$= \frac{TR + tr}{TtRr + Ttrr + ttRr + ttrr}$$

or

$$= \frac{Tr + tR}{TtRr + Ttrr + rrRr + ttrr}$$

Linkage can also be detected, but less efficiently, by selfing the F$_1$ and observing the segregation ratio of the F$_2$ to determine if there is any deviation from the expectation, based on independent assortment (e.g. 1 *TTRR*:2 *TTRr*:1 *TTrr*:2 *TtRR*:4 *TtRr*:2 *Ttrr*:1 *ttRR*:2 *ttRr*:1 *ttrr*).

The relative positions of three loci can be *mapped by considering frequency of progeny from a* ***three gene test cross***.

To examine three gene test crosses consider the following cross between two parents where one parent has the genotype *AABBCC* and the other has the genotype *aabbcc*. The F$_1$ would have the genotype *AaBbCc*. In order to perform the test cross it is necessary to cross the F$_1$ family with a completely recessive genotype (i.e. *aabbcc*, in this case the recessive parent) and observe the frequency of the eight possible phenotypes. In our example the frequencies are as shown in Table 5.3.

If all three loci segregated independently, for example they are on different chromosomes, we would expect the phenotype frequencies of the eight genotypes to be

Table 5.3 Observed and expected genotypes obtained from backcrossing the F$_1$ progeny from a cross between two parents where one parent has the genotype *AABBCC* and the other has the genotype *aabbcc*, to the recessive (*aabbcc*) parent.

Phenotype	Observed	Expected
*A_B_C_*	500	149.375
aabbcc	510	149.375
*aaB_C_*	50	149.375
A_bbcc	55	149.375
*A_bbC_*	35	149.375
aaB_cc	38	149.375
A_B_cc	4	149.375
*A_bbC_*	3	149.375
Total	1195	

the same at 149.375 (or the total number of observations divided by the expected number of phenotype classes, 1195/8 = 149.375). Obviously, the observed frequencies are very different from those expected.

The first point to note is that the frequency of the parental genotypes is considerably higher than expected and all other **recombinant** types are less than expected.

It is necessary to know the relative order that the loci appear on the chromosome. The middle locus can usually be determined from the genotype that is observed at the lowest frequency. In this example the lowest observed phenotypes are *A_B_cc* and *aabbC_*, both of which are effectively parental types at the *A* and *B* locus but involves non-parental combinations with *C*. As recombination of only the centre gene will involve a **double cross over** event the frequency of occurrence will be lowest. Conversely, the pair with most phenotypic observed classes for recombinants (not counting double crossovers) is the outside genes. From this, the *C*-locus would appear to be the middle one.

Consider the possible combinations of alleles and the frequency that they occur.

For the *A − B* loci we have:

Parental types	*A_B_*	500 + 4	= 504
	aabb	510 + 3	= 513
Recombinants	*aaB_*	50 + 38	= 88
	A_bb	55 + 35	= 90

So percentage of recombinants = (178/1195) × 100 = 14.9%

For *A–C* loci we have:

Parental types	*A_C_*	500	+ 35 = 535	
	aacc	510	+ 38 = 548	
Recombinants	*aaC_*	50	+ 3 = 53	
	A_cc	55	+ 4 = 59	

So percentage of recombinants = (112/1195) × 100 = 9.4%.

For the *B–C* loci we have:

Parental types	*B_C_*	500	+ 50 = 550	
	bbcc	510	+ 55 = 565	
Recombinants	*bbC_*	35	+ 3 = 38	
	B_cc	38	+ 4 = 42	

So percentage of recombinants = (80/1195) × 100 = 6.7%.

We obtain the ***map distance*** as being equivalent to the percentage recombinants and so we have:

A- - - - - - - - - - - -14.9 mμ- - - - - - - - - - - - -B

A- - - - - -9.4 mμ- - - - - - -C- - -6.7 *mμ*- - -B

From the map, it is clear that the map distance *A* to *B* (14.9 mμ) is less than the added distance A–C plus C–B. The method of calculation used to estimate these distances assumes that only the non-parental types are included, as would be the case where the linkage between two loci is considered. Where three loci are involved it is possible to include the double cross over recombination events in estimating the distance between the furthest two loci. In this case we could calculate the *A–B* distance from:

Non recombinant types	*A_B_C_*	500	= 500
	aabb	510	= 510
Recombinant events	*aaB_*	50 + 38	= 88
	A_bb	55 + 35	= 90
Twice double cross over recombinants		2(3 + 4)	= 14

So percentage of recombinants = (192/1195) × 100 = 16.1%, which is the same distance that would be estimated by simple adding *A–C* to *C–B*.

This indicates a general flaw in linkage map distance estimation in that where there is no 'centre gene' locus,

it is impossible to detect double recombination events, and as a result map distances are always going to provide under estimates of actual recombination frequencies.

To avoid such discrepancies, recombination frequencies, which are not additive, are usually converted to a cM scale using the function published by J.B.S. Haldane in 1919, being:

$$cM = (-\ln(1 - 2R)) \times 50$$

where *R* is the recombination frequency. From this we see that the map distances transform to:

A- - - - - - - - - - - -17.6 *cM*- - - - - - - - - - -B

A- - - - - -10.4 *cM*- - - - -C- - -7.2 *cM*- - -B

The standard error of these map distances can be calculated using the equation:

$$s.e. = [R/(1 - R)]/n$$

where *R* is the recombination frequency, and n is the number of plants observed in the three-way test cross.

Pleiotropy

Very tight linkage between two loci can be confused with pleiotropy, the control of two or more characters by a single gene. For example, the linkage of resistance to the soybean cyst nematode and seed coat colour seems to be a case of pleiotropy because the two characters were always inherited together. The only way linkage and pleiotropy can be distinguished, is effectively the negative way, that is, to find a crossover product, such as a progeny homozygous for resistance and yellow seed coat. Resistance and yellow seed coat could never occur in a true breeding individual if true pleiotropy was present. Thousands of individuals may have to be grown to break a tight linkage that appears to be one of pleiotropy.

Epistasis

Different loci may be independent of each other in their segregation and recombination patterns. Independence of gene transmission, however, does not necessarily imply independence of gene action or expression. In fact, in terms of its final expression in the phenotype of the individual, ***no gene acts by itself.***

A character can be controlled by genes that are inherited independently but that interact to form the final phenotype. The interaction of genes at different loci that affect the same character is called either non-allelic interaction or *epistasis*. Epistasis was originally used to describe two different genes that affect the same character, one that masks the expression of the other. The gene that masks the other is said to be epistatic to it. The gene that is masked was termed hypostatic. Epistasis causes deviations from the common phenotypic ratios in F_2 such as 9 : 3 : 3 : 1 that indicates segregation of two independent genes, each with complete dominance. Phenotypic ratios in F_2 for two unlinked genes as influenced by degree of dominance at each locus and epistasis between loci are shown in Table 5.4.

In some ways the interaction between two different loci, in which the allele at one locus affects the expression of the alleles at another, will remind you of the phenomenon of dominance. The two phenomena are essentially different, however. **Dominance** always refers to the expression of one member of a pair of alleles relative to the other at the same locus, as opposed to another locus; **epistasis** is the term generally used to describe effects of non-allelic genes on each other's expression, in other words their interaction.

Table 5.4 Phenotypic ratios of progeny in the F_2 generation for two unlinked genes (where *A* is dominant to *a*, and *B* is dominant to *b*), and epistasis between loci.

F$_2$ phenotype				Genetic explanation
*A_B_*	*A_bb*	*aaB_*	*aabb*	
9	3	3	1	No epistasis
9	3	4	0	Recessive epistasis: *aa* epistatic to *B* and *b*
12	0	3	1	Dominant epistasis: *A* epistatic to *B*, or *b*
13	0	3	0	Dominant and recessive epistasis: *A* epistatic to *B* and *b*; *bb* epistatic to *A* and *a*. *A_* and *bb* produce identical phenotypes
9	0	7	0	Duplicate recessive epistasis: *aa* epistatic to *B*, and *b*; and *bb* epistatic to *A* and *aa*
15	0	0	1	Duplicate dominant epistasis: *A* epistatic to *B* and *b*; *B* epistatic to *A* and *aa*

Qualitative inheritance in tetraploid species

Compared to crop species that are cultivated as pure inbred lines, there has been comparatively little research carried out on the inheritance in auto-tetraploid species. This has been primarily due to the fact that the major autotetraploid crop species (e.g. potato) are clonally reproduced or are outbreeding species that suffer severe inbreeding depression (i.e. alfalfa). Many (or all) of these cultivars are highly heterozygous and hence it is not as easy to carry out simple genetic experiments.

Major gene inheritance in auto-tetraploids has been the topic of many research/breeding groups with the aim of parental development. Consider for example the potato crop, where there are several single gene traits that control resistance to Potato Virus X, Potato Virus Y, Potato Cyst Nematode (*G. rostochiensis*) and late blight (*Phytophthora infestans*). All these qualitative traits show complete dominance. The technique used to develop parents in a breeding programme is aimed at increasing the proportion of desirable offspring in sexual crosses and in the extreme to avoid the need to test breeding lines for the presence of the allele of interest. The technique is called *multiplex breeding*.

In tetraploid crops any genotype may be *nulliplex* (*aaaa*), having no copies of the desired allele at the specific (*A*) locus; *simplex* (*Aaaa*), having only one copy of the desirable resistance allele (*A*); *duplex* (*AAaa*), having two copies of the allele; *triplex* (*AAAa*), having three copies of the allele or *quadruplex* (*AAAA*) having four copies of the allele (i.e. homozygous at that locus). If the alleles show dominance then genotypes which have at least one copy of the gene will, phenotypically, appear identical in terms of their resistance. But they will differ in their effectiveness as parents in a breeding programme. To determine the genotype of a clonal line, test crossing is necessary.

To illustrate the usefulness of multiplex breeding in potato, consider the problem of developing a parental line which, when crossed to any other line (irrespective of the genotype of the second parent) will give progeny all of which will be resistant to potato cyst nematode by having at least one copy of the H$_1$ gene (a qualitative resistance gene conferring resistance to all UK populations of the damaging nematode *Globodera rostochiensis* and which has been shown to give relatively durable resistance).

The aim of multiplex breeding is therefore to develop parental lines which are either triplex or quadruplex (three or four copies of the desirable allele, respectively) for the H_1 allele. When crossed to any other parent these multiplex lines will produce progeny that have at least a single copy of the H_1 gene and due to dominance, all will be phenotypically resistant. It will, therefore, not be necessary to screen for resistance in these progeny. The ratios of resistant to susceptible amongst the progeny of genotypes derived by crossing a simplex, duplex, triplex and quadruplex to a nulliplex are shown in Table 5.5.

Genotype ratios from all possible cross combinations between nulliplex, simplex, duplex, triplex or quadruplex (Table 5.6)

It was at first thought that developing multiplex parents would be very difficult, instead it proved simply to be a matter of effort and application. The main difficulty lies in the fact that progeny tests need to be carried out to test the genetic make-up of the dominant parental lines (very similar to backcrossing where the non-recurrent parent has a single recessive gene of

interest). Consider one of the worst situations, where both starting parents are simplex. Three quarters of the progeny will be resistant, but only one quarter will be duplex. So one quarter of the progeny will be nulliplex, and hence susceptible and so on testing can be immediately discarded. The progeny need, however, to be testcrossed to a nulliplex to distinguish the duplex from simplex genotypes in the progeny. Once identified, the selected duplex lines are inter-crossed or selfed. From their progeny 1/36 will be quadruplex and 8/36 will be triplex. A single round test cross will be necessary to distinguish the triplex and quadruplex lines from those progeny that are either duplex or simplex. A second round test cross will be necessary (a second backcross to a nulliplex lines) in order to distinguish the quadruplex from triplex lines.

Once quadruplex lines have been identified, these can be continually inter-crossed, or selfed, without further need to test. Similarly, a quadruplex or triplex parental lines can be used in cross combination with any other parental line and 100% of the resulting progeny will be

Table 5.5 The ratios of resistant to susceptible progeny amongst the genotypes derived by crossing a simplex, duplex, triplex and quadruplex resistant gene parent to a nulliplex parent.

Cross type	Phenotype Resistant : Susceptible	% Resistant in progeny
Simplex (*Rrrr*) × nulliplex (*rrrr*)	1 : 1	50
Duplex (*RRrr*) × nulliplex (*rrrr*)	5 : 1	83
Triplex (*RRRr*) × nulliplex (*rrrr*)	1 : 0	100
Quadruplex (*RRRR*) × nulliplex (*rrrr*)	1 : 0	100

Table 5.6 Genotype ratios from all possible cross combinations between nulliplex, simplex, duplex, triplex or quadruplex parents.

Cross	Nulliplex (N)	Simplex (S)	Duplex (D)	Triplex (T)	Quadruplex (Q)
Nulliplex (*aaaa*)	All N				
Simplex (*Aaaa*)	1S : 1N	1D : 2S : 1N			
Duplex (*AAaa*)	1D : 4S : 1N	1T : 5D : 5S : 1N	1Q : 8T : 18D : 8S : 1N		
Triplex (*AAAa*)	1D : 1S	1T : 2D : 1S	1Q : 5T : 5D : 1S	1Q : 2T : 1D	
Quadruplex (*AAAA*)	All D	1T : 1D	1Q : 4T : 1D	1Q : 1T	All Q

resistant. Multiplex breeding can be used in a similar way to develop parents in hybrid cross combinations.

The chi-square test

If plant breeders have an understanding of the inheritance of simply inherited characters it is possible to predict the frequency of desired genes and genotypes in a breeding population. Breeders are often asking questions relating to the nature of inheritance as well as the number of alleles or loci involved in the inheritance of a particular character of importance. For example, it is often valuable to determine whether a single gene is dominant or additive in inheritance. In the dominant case the F_2 family will segregate in a $3:1$ dominant : recessive ratio, while in the latter a $1:2:1$ homozygous dominant : heterozygous : homozygous recessive ratio.

In segregating families such as those noted above, the ratios actually observed will not be exactly as predicted due to sampling error. For example a coined tossed 10 times does not always result in 5 heads and 5 tails. So the breeder is faced with interpreting the ratios that are observed in terms of what is expected, so it might be necessary to decode if a particular ratio is $1:2$ or $1:3$ or $3:4$ or $9:7$. How can this be done objectively? The answer is to use chi-square tests.

It should be clear that the significance of a given deviation is related to the size of the sample. If we expect a $1:1$ ratio in a test involving six individuals, an observed ratio of $4:2$ is not at all bad. But if the test involves 600 individuals, an observed ratio of $400:200$ is clearly a long way off. Similarly, if we test 40 individuals and find a deviation of 10 in each class, this deviation seems serious:

observed	30	10
expected	20	20
obs − exp	10	10

But if we test 200 individuals, the same numerical deviation seems reasonably enough explained as a purely chance effect:

observed	90	110
expected	100	100
obs − exp	10	10

The statistical test most commonly used when such a problems arises, is simple in design and application. Each deviation is *squared,* and the expected number in its class then divides each squared deviation. The resulting quotients are then all added together to give a single value, called the chi-square (χ^2). To substitute symbols for words, let d represent the respective deviations (observed minus expected), e the corresponding expected values, then:

$$\chi^2 = \sum (d^2/e)$$

We can calculate chi-square for the two arbitrary examples above to show how this value relates the magnitude of the deviation to the size of the sample.

	Sample of 40 individuals		Sample of 200 individuals	
Observed (obs)	30	10	90	110
Expected (exp)	20	20	100	100
obs − exp(d)	10	10	10	10
d^2	100	100	100	100
d^2/exp	5	5	1	1
χ^2		10		2

You will note that the value of chi-square is much larger for the smaller population, even though deviations in the two populations are numerically the same. In view of our earlier common-sense comparison of the two, this is a practical demonstration that the calculated value of chi-square is related to the significance of a deviation. It has the virtue of reducing many different samples, of different sizes and with different numerical deviations, to a common scale for comparison.

The chi-square test can also be applied to samples including more than two classes. For example, the table below shows the chi-square analyses of the tall, 6-row × short 2-row barley example earlier. Suppose that a test cross was carried out where the heterozygous F_1 (*TtSs*) is crossed to a genotype with the recessive alleles at these loci. When this was actually done, and 400 test cross progeny were grown out, there were 96 tall and 2-row, 107 tall and 6-row, 97 short and 6-row, and 102 short and 2-row. Inspection of this would show that there are a higher proportion of the original parental types (tall/6-row and short/2-two) than the recombinant types. The question therefore, is this linkage, or simple random sampling variation. To determine this we would use the χ^2 test.

	Tall, 6-row	Tall, 2-row	Short, 6-row	Short, 2-row
Observed (obs)	112	89	93	106
Expected (exp)	100	100	100	100
obs − exp(d)	12	11	7	6
d^2		144	121	49	36
d^2/exp		1.44	1.21	0.49	0.36

$$\chi^2_{3df} = \sum(d^2/\text{exp}) = 3.50 \text{ n.s.}$$

The number of degrees of freedom in tests of genetic ratios is almost always *one less than the number of classes*. To be more precise, it is the number of observable data that are independent. For example, in a two gene test cross there are four possible phenotypes, expected in equal frequency. If 400 plants are observed and there are 50 in the first group, 120 in the second group and 140 in the third group; then by definition there must be 90 is the last group $(400 - 50 - 120 - 140 = 90)$. A test of $1:2:1$ ratio would have two degrees of freedom. In just the same way, if we were testing two groups in tests of $1:1$ or $3:1$ ratios there is one degree of freedom.

Do not confuse assigning degrees of freedom to genetic frequencies with degrees of freedom in χ^2 contingency tables. In a two-way contingency table, with pre-assigned row and column total, one value can be filled arbitrarily, but the others are then fixed by the fact that the total must add up to the precise number of observations involved in that row or column. When there are four classes, any three are usually free, but the fourth is fixed. Thus, when there are four classes, there are usually three degrees of freedom.

Remembering the example given earlier, calculated chi-square $= 10$ for one degree of freedom for the sample of 40 individuals. We can look this value up in the probability tables for chi-square and in this case chance alone would be expected in considerably less than one in a hundred independent trials to produce as large a deviation as that obtained. We cannot reasonably

accept chance alone as being responsible for this particular deviation; it represents an event that would occur, on a chance basis, much less often than the one-time-in-twenty that we have agreed on as our point of rejection; this event would occur less often than even the one-time-in-a-hundred that we decided to regard as highly significant.

In the case of the example we noted for barley, the expected frequency of each of the 4 phenotypes if no linkage is present would be 100 (total of 400, with four equal expectations), which would lead to deviations of 12, 11, 7 and 6, these squared and divided by the expected value $(144/100 + 121/100 + 49/100 + 36/100)$ gives a χ^2 value of 3.5. This value is compared in probability tables for χ^2 values and it falls just below the 50% probability table value with 3 degrees of freedom, clearly not close to the accepted 5% probability we accept as showing significance. We therefore say that there is no evidence that linkage exists and that the observed deviations are likely to have happened by random chance (sampling error).

Improper use of chi-square

The two most important reservations regarding the straightforward use of the chi-square method in genetics are:

- Chi-square can usually be applied only to numerical frequencies themselves, not to percentages or ratios derived from the frequencies. For example, if in an experiment one expects equal numbers in each of two classes, but observes 8 in one class and 12 in the other, we might express the observed numbers as 40% and 60%, and the expected as 50% in each class. A chi-square value computed from these percentages can not be used directly for the determination of p. When the classes are large, a chi-square value computed from percentages can be used, if it is first multiplied by $n/100$, where n is the total number of individuals observed.

- Chi-square cannot properly be applied to distributions in which the expected frequency of any class is less than 5. In fact, some statisticians suggest that a particular correction **be** applied if the frequency of any class is less than 50. However, the approximations involved in chi-squares are close enough for most practical purposes when there are more than 5 expected in each class.

Family size necessary in qualitative genetic studies

It is usual in genetic work for the scope of the experiments to be limited by such considerations as lack of available space, labour or money, etc. It is therefore essential to make the best use of available resources, which will be a function of the number of plants/plots that need to be raised. Achieving this usually requires considerable care and planning of experiments. Often statistical considerations can be of great value.

In many experiments it is desirable to be able to pick out certain genotypes, usually (but not always) homozygous, with the aim of developing superior cultivars. It may also be necessary to detect some non-conformity. For example to detect any linkage effects will involve test crossing F_1 lines onto a recessive parent. It would be advantageous to keep the population size down to a minimum while also assuring with high probability that the experiment will be sufficiently large so as to detect differences required.

Consider now the question of detecting homozygotes in a segregating population. Any progeny which fails to segregate could have derived from a homozygous parent or could fail to show segregation because of sampling error. The greater the numbers that are examined, the lower the probability that sampling error will interfere with the interpretation of experimental results. The minimum size of progeny designed to test a particular individual is then a statistical question involving consideration of the probability that any individual, in a family derived from a heterozygote, will be of the recessive type, and also the permissible maximum probability of obtaining a misleading result.

Consider the following example. Suppose you wish to test a series of individuals phenotypically dominant for a single gene, in order to identify the homozygous individuals, by using a test cross to a genotype being homozygous recessive. The progeny of the homozygotes will not show segregation while progeny from the heterozygotes will segregate in a 1 : 1 ratio (i.e. $1/2Aa : 1/2aa$). The main error that will arise will be from our failure to detect any segregants in the progenies from crosses that actually do have a heterozygous parent. Let us also assume that we do not want the test to fail with greater probability than once in 100 experiments or test (i.e. $p < 0.01$). In the progeny of a heterozygote each individual has a chance of 1/2 of

containing a dominant allele. Then a family of n will be expected to have $(1/2)^n$ individuals with one dominant allele. Therefore we can predict the possibility of having no such individuals, which represents the misleading (error) result which must be avoided and must not occur at higher frequency than 1 in 100. Then the minimum value of n is given by the solution of the equation:

$$1 - (1/2)^n = 1/100$$

Taking natural logarithms this becomes:

$$n \times \ln(1 - 1/2) = \ln(1/100)$$

Therefore:

$$n = \ln(1/100)/\ln(1 - 1/2)$$

In this example $n = 2/0.3010 = 6.6$. Therefore the smallest family size needed would be seven or more, in order to be 99% certain of detecting the difference.

This leads to the generalized formula:

$$n = \ln(1 - p)/\ln(1 - x)$$

where n is the number that needs to be evaluated, p is the probability of having at least one type of interest (i.e. 90% certain $= p = 0.9$) and x is the frequency of the desirable genotype/phenotype etc.

It is often necessary in plant breeding to estimate the number of individuals that need to be grown or screened in order to identify at least r individuals (where r is greater than one). It is never a good idea to simply multiply n (from above), the number that needs to be grown to ensure at least one by r the number required, as this will result in a gross overestimation.

An alternative is to use an extension to the above equation. The mathematics behind this equation is outside the scope of this book. However, the equation itself can be useful. It is:

$$n = ([2(r - 0.5) + z^2(1 - q)]$$
$$+ z[z^2(1 - q)^2 + 4(1 - q)(r - 0.5)]^{0.5})(2q)^{-1}$$

where n = total number of plants that need to be grown, r = the number of plants with the required genes that need to be recovered; q = the frequency of plants with the desired genes; p = the probability of recovering the desired number of plants with the desired genes and z = a cumulative normal distribution frequency distribution (area under standardized normal curve from

0 to z) a function of probability (p). For the sake of simplicity, $z = 1.645$ for $p = 0.95$ (i.e. 95% certain) and $z = 2.326$ (for 99% certainty).

For example, consider the frequency of homozygous recessive genotypes (*aabb*) resulting from the cross *AAbb* × *aaBB* at the F$_3$ stage. How many F$_3$ lines would need to be evaluated to be 95% certain of obtaining 10 homozygous recessive genotypes? Here the probability of *aabb* is 9/64 (i.e. $q = 0.141$), $p = 0.95$, therefore $z = 1.645$.

$$n = ([2(10 - 0.5) + 2.706(1 - 0.141)]$$
$$+ 1.645[(2.706(1 - 0.141)^2 + 4(1 - 0.141)$$
$$\times (10 - 0.5)]^{0.5})(2 \times 0.141)^{-1}$$
$$= 110.69$$

Therefore 111 F$_3$ lines need to be grown.

QUANTITATIVE GENETICS

The basis of continuous variation

With qualitative inheritance, the segregating individual phenotypes are usually easily distinguished and fall into a few phenotypic classes (i.e. tall or short). Characters that are controlled by multiple genes do not fall into such simple classifications. Consider a hectare field of potatoes, all planted with one cultivar and so every single plant in the field should be of identical genotype (ignoring mutation and errors). In that field there are likely to be 11 000 plants. At harvest, you have been given the task of harvesting all 11 000 plants and weighing the tubers that come from each plant separately.

Would you expect all the potato plants to have exactly the same weight of potatoes? – probably not. Indeed the weights can be presented in the form of a histogram (Figure 5.1) where 11 000 yields were divided into 17 weight classes. The variation in weight is obvious; some plants produce less than 0.5 kg of tubers, while others produce over 5 kg. Most, however, are grouped around the average of 3.2 kg weight.

Yield in potato, as in other crops, is polygenically inherited. Yield is therefore not controlled by a single gene but by many genes, all acting collectively. Thus yield has more chance of some of the processes these many genes control being affected by differences in the 'environment'. As the example above involved harvesting plants which are all indeed clones of the same genotype, then the variation observed is due entirely to the environmental conditions that each plant was subjected to.

One of the major differences between single gene inheritance and multiple gene inheritance is that the former is less affected or influenced by the environmental conditions compared to the later. When potato yield are being recorded we are recording the expression of the phenotype, but are interested in the genotype. The two are related by the equation:

$$P = G + E + GE + \sigma_e^2$$

where P is the phenotype, G is the genotypic effect, E is the effect of the environment in which the genotype is grown, GE is the interaction between the genotype and the environment and σ_e^2 is a random error term.

Single gene characters are often less affected by environment and usually do not show much influence

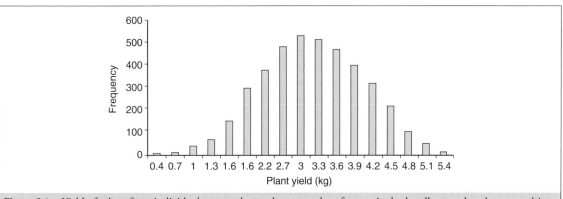

Figure 5.1   Yield of tubers from individual potato plants taken at random from a single clonally reproduced potato cultivar.

of genotype × environment interactions (i.e. the difference in expression of the two alleles is large in comparison to the variation caused by environmental changes). For example, a potato genotype with white flowers (a qualitatively inherited trait) will always have white flowers in any environment in which the plants produce flowers. Conversely, quantitatively inherited traits are greatly influenced by environmental conditions and genotype × environment interactions are common, and can be large. The greatest difficulty plant breeders face is dealing with quantitative traits and in particular in deciding on the better genotypes based on their phenotypic performance. This is why it is critically important to employ appropriate experimental design techniques to genetic experiments and also plant breeding programmes.

A major part of quantitative genetics research related to plant breeding has been directed towards partitioning the variation that is observed (i.e. phenotypic variation) into its genetic and non-genetic portions. Once achieved this can be taken further to further divide the genetic portion into that which is additive in nature and that which is non-additive (quite often dominance variation). Obviously in breeding self-pollinating crops, the additive genetic variance is of primary importance since it is that portion of the variation due to homozygous gene combinations in the population and is what the breeder is trying to obtain. On the other hand, variance due to dominance is related to the degree of heterozygosity in the population and will be reduced (to zero) over time with inbreeding as breeding lines move towards homozygosity.

Let us now return to the potato weights as one, of many possible examples of continuous variation. If the frequency distribution of potato yields is inspected there are two points to note: (1) the distribution is symmetrical (i.e. there are as many high yield as really low yields); and (2) the majority of potato yield were clustered around a weight in the middle. As we have taken a class interval of 0.3 kg to produce this distribution, the figure does not look particularly continuous. However, we know that potato yields do not go up in increments of 0.3 kg but show a more continuous and gradual range of variation. If we use more class intervals in this example we will produce a smoother histogram, and if we use an infinity small class intervals it will result in a continuous bell-shaped curve. The shape of this curve is highly indicative of many aspects of plant science because it is

a distribution called a *normal distribution*, and occurs in a wide variety of aspects relating to plant growth, and particularly to quantitative genetics.

Describing continuous variation

The normal distribution

The 11,000 potato plant weights discussed above are a sample, albeit a large sample, of possible potato weights from individual plants. It is possible to predict mathematically the frequency distribution for the population as a whole (i.e. every possible potato plant of that cultivar grown), provided it is assumed that the sample is representative of the population (i.e. that our sample is an unbiased sample of all that was possible).

It is not necessary to actually draw normal distributions (which, even with the aid of computer graphics, are difficult to do accurately). Most of the properties of a normal distribution can be characterized by two statistics, the mean or average (μ) of the distribution and the standard deviation (σ), a measure of the '*spread*' of the distribution.

There are in fact two means, the mean of the sample and the mean of the population from which the sample was drawn. The latter is represented by the symbol μ, and can, in reality, seldom be known precisely. The *sample mean* is represented by $\bar{x}$ (spoken x bar), and it can be known with complete accuracy. The best estimate of a *population mean* (μ) is generally the actual mean of an unbiased sample drawn from it ($\bar{x}$). The population mean is thus best estimated as:

$$\mu = \bar{x} = (x_1 + x_2 + x_3 + \cdots + x_n)/n = \sum_{i}^{n}(x_i)/n$$

where $\sum_{i}^{n}(x_i)$ is the sum of all x values from $i=1$ to n.

The standard deviation is an ideal statistic to examine the variation that exists within a data set. For any normal distribution, approximately 68% of the population sampled will be within one standard deviation from the mean, approximately 95% will be within two standard deviations (Figure 5.2), and approximately 99% will be within three standard deviations of the mean.

Once again, it is necessary to distinguish between the actual standard deviation of a population, all of whose members have been measured, or of a particular sample, and the estimated standard deviation of a population based on measuring a sample of individuals from it. The former is represented by the symbol σ (Greek letter

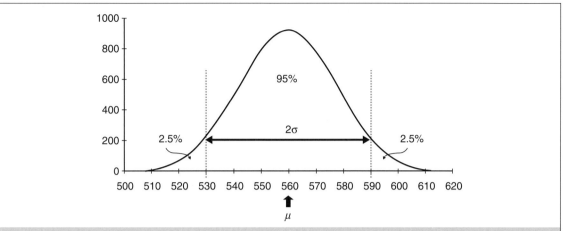

Figure 5.2 95% of a population which is normally distributed will lie within one standard deviation from the population mean.

sigma) and is defined thus:

$$\sigma = \sqrt{\left[\sum_{i=1}^{n}(x_i - \mu)^2/n\right]}$$

and the latter is represented by the symbol $\hat{s}$, where:

$$\hat{s} = \sqrt{\left[\sum_{i=1}^{n}(x_i - \bar{x})^2/(n-1)\right]}$$

Another measure of the spread of data around the **mean** is the *variance*, which is the square of the standard deviation. The *estimated variance* of a population is given by:

$$\hat{s}^2 = \left[\sum_{i=1}^{n}(x_i - \bar{x})^2\right]/(n-1)$$

and the *actual variance* of a sample drawn from a population (or of an entire population if every member of it has been measured) is given by:

$$\sigma^2 = \left[\sum_{i=1}^{n}(x_i - \mu)^2\right]/n$$

Calculators are often programmed to give means and either standard deviations or variances with a few key strokes once data have been entered. However, in case it is necessary to derive these descriptive statistics semi-manually, it is useful to know about alternative

equations for s^2 and σ^2:

$$\hat{s}^2 = \left\{\sum_{i=1}^{n}(x_i^2) - \left[\left[\sum_{i=1}^{n}(x_i)\right]^2/n\right]\right\}/(n-1)$$

$$\sigma^2 = \left\{\sum_{i=1}^{n}(x_i^2) - \left[\left[\sum_{i=1}^{n}(x_i)\right]^2/n\right]\right\}/n$$

Although these *look* more complicated than those given previously, they are *easier to use* because the mean does not have to be worked out first (which would entail entering all the data into the calculator twice). Note the difference between $\sum(x_i^2)$ (each value of x squared and then the squares totalled) and $[\sum(x_i)]^2$ (the values of x totalled and then the sum squared).

Standard deviations, as measures of spread around the mean, are probably intuitively more understandable than variances, for example, 68% of the population fall within one standard deviation of the mean. Why introduce the complication of variances? Well, *variances are additive* in a way standard deviations are not. Thus, if the variances attributable to a variety of factors have been estimated, it is mathematically valid to sum them to estimate the variance due to all the factors acting together. Similarly, a total *variance* can be *partitioned* into the variances attributable to a variety of individual factors. These operations, which are used extensively in quantitative genetics, cannot so readily be performed with standard deviations.

Variation between data sets

Two basic procedures are frequently used in quantitative genetics to interpret the variation and relationship that exists between characters, or between one character evaluated in different environments. These are simple linear regression and correlation.

A straight line regression can be adequately described by two estimates, the slope or *gradient* of the line (**b**) and the *intercept* on the y-axis (**a**). These are related by the equation:

$$y = bx + a$$

It can be seen that **b** is the gradient of the line, because a change of one unit on the x-axis results in a change of **b** units on the y-axis. If x and y both increase (or both decrease) together, the gradient is positive. If, however, x increases while y decreases or vice versa, then the gradient is negative. When $x = 0$, the equation for y reduces to:

$$y = a$$

and **a** is therefore the point at which the regression line crosses the y-axis. This intercept value may be equal to, greater than or less than zero.

The formulation and theory behind regression analysis will not be described here and are not within the scope of this book. However, the gradient of the best fitting straight line (also known as the *regression coefficient*) for a collection of points whose coordinates are x and y is estimated as:

$$b = [SP(x, y)/SS(x)]$$

where, $SP(x, y)$ is the *sum of products* of the deviations of x and y from their respective means ($\bar{x}$ and $\bar{y}$) and $SS(x)$ is the *sum of the squared deviations* of x from its mean. It will be useful to have an understanding of the regression analysis and to remember the basic regression equations.

Now, $SP(x, y)$ is given by the equation:

$$SP(x, y) = \sum_{i=1}^{n}(x_i - \bar{x})(y_i - \bar{y})$$

although in practice it is usually easier to calculate it using the equation:

$$SP(x, y) = \sum_{i=1}^{n}(x_i y_i) - \left[\sum_{i=1}^{n}(x_i)\sum_{i=1}^{n}(y_i)\right]\Big/ n$$

The comparable equations for $SS(x)$ are:

$$SS(x) = \sum_{i=1}^{n}(x_i - \bar{x})^2$$

$$SS(x) = \sum_{i=1}^{n}(x_i^2) - \left\{\left[\sum_{i=1}^{n}(x_i)\right]^2\right\}\Big/ n$$

Notice that a sum of squares is really a special case of a sum of products. You should also note that if every y value is exactly equal to every x value, then the equation used to estimate **b** becomes, $SS(x)/SS(x) = 1$.

Having determined **b**, the intercept value is found by substituting the mean values of x and y into the rearranged equation.

$$a = \bar{y} - b\bar{x}$$

In regression analysis it is always assumed that one character is the dependant variable and the other is independent. For example, it is common to compare parental performance with progeny performance (see Chapter 6) and in this case then progeny performance would be considered the dependant variable and parental performance independent. The performance of progeny is obviously dependent on the performance of their parents, and not *vice versa*.

The degree of association between any two, or a number of different characters can be examined statistically by the use of *correlation analysis*. Correlation analysis is similar in many ways to simple regression but in correlations there is no need to assign one set of values to be the *dependant variable* while the other is said to be the *independent variable*. Correlation coefficients (r) are calculated from the equation:

$$r = \frac{SP(x, y)}{\sqrt{[SS(x) \times SS(y)]}}$$

where $SP(x, y)$ is again the sum of products between the two variables, $SS(x)$ is the sum of squares of one variable (x) and $SS(y)$ is the sum of squares of the second variable (y), and:

$$SP(x, y) = \left[\sum_{i=1}^{n}(x_i - \bar{x})(y_i - \bar{y})\right]\Big/(n - 1)$$

$$SS(x) = \left[\sum_{i=1}^{n}(x_i - \bar{x})^2\right]\Big/(n - 1)$$

$$SS(y) = \left[\sum_{i=1}^{n}(y_i - \bar{y})^2\right]\Big/(n - 1)$$

These can of course, sometimes be calculated more easily by:

$$SP(x, y) = \left\{ \sum_{i=1}^{n} x_i y_i - \left[\sum_{i=1}^{n} x_i \sum_{i=1}^{n} y_i \right] \Big/ n \right\} \Big/ (n-1)$$

$$SS(x) = \left\{ \sum_{i=1}^{n} x_i^2 - \left[\sum_{i=1}^{n} x_i \right]^2 \Big/ n \right\} \Big/ (n-1)$$

$$SS(y) = \left\{ \sum_{i=1}^{n} y_i^2 - \left[\sum_{i=1}^{n} y_i \right]^2 \Big/ n \right\} \Big/ (n-1)$$

Correlation coefficients (r) range in value from -1 to $+1$. r values approaching $+1$ show very good positive association between two sets of data (i.e. high values for one variable are always associated with high values of the other). In this case, we say that the two variables are **positively correlated**. Values of r which are near to -1 show disassociation between two sets of data (i.e. a high value for one variate is always associated with a low value in another). In this case we say that the two variables are **negatively correlated**. Values of r that are near to zero indicate that there is no association between the variables. In this case a high value for one variable can be associated with a high, medium or low value of the other.

Relating quantitative genetics and the normal distribution

Consider two homozygous canola (*Brassica napus*) cultivars (P_1 and P_2). The yield potential of P_1 is 620 kg/plot, and is higher than P_2, which has a yield potential of 500 kg/plot. When these two cultivars were crossed and the F_1 produced, the yield of the F_1 progeny was exactly midway between both parents (560 kg/plot). This would suggest that additive genetics effects rather than dominance were present.

If yield in canola were controlled by a single locus and two alleles (which it is not), we would have:

$$P_1 \times P_2$$
$$AA \times aa$$

It should be noted here that upper and lower case letter denoting alleles do not signify dominance as in qualitative inheritance, but rather differentiate between alleles. It is common to assign uppercase letters to alleles from the parent with the greater expression of the trait, always donated as P_1.

There are two alleles involved. Assume that the uppercase alleles add 60 kg/plot to the base performance of a plant, and lowercase alleles add nothing. In this case the base performance is equal to $P_2 = 500$ kg/plot. Therefore, $P_1 = AA = 620$ kg/plant ($500 + 60 + 60$), $P_2 = aa = 500$ kg/plot ($500 + 0 + 0$), The $F_1 = Aa = 560$ kg/plot ($500 + 60 + 0$). At F_2 we have a ratio of 1 *AA* : 2 *Aa* : 1 *aa*, and we would have three types of plants in the population: *AA* = 620 kg/plot; *Aa* = 560 kg/plot and *aa* = 500 kg/plot.

Obviously, yield in canola is not controlled by one gene. However, let us progress gradually and assume that two loci each with two alleles are involved. We now have:

$$P_1 \times P_2$$
$$AABB \times aabb$$

In this case (assuming alleles at different loci have equal effect) each of the two uppercase alleles would each add 30 kg/plot to the base weight. The $F_1 = AaBb = 560$ kg/plot (the same as if only one gene was involved). However at the F_2 we have 16 possible allele combinations that can be grouped according to number of uppercase alleles (or yield potential).

			AAbb	
			AaBb	
		aabB	*aABa*	*AABb*
		aaBb	*AabB*	*AAbB*
		aAbb	*aAbB*	*AaBB*
	aabb	*Aabb*	*aaBB*	*aABB*	*AABB*
	500	530	560	590	620

Extending in the same manner one more time we see that the frequency distribution of the phenotypic classes in the F_2 generation when *three* genes having equal additive effects, and which segregate independently, are:

```
                        aaBbCC
                        aAbBcC
                        aAbBCc
                        aAbbCC
                        AabbCC
            aabbCC  AabBcC  AABBcc
            aabBcC  AabBCc  AABbCc
            aabBcc  AabbCc  AABbcC
            aaBbcC  AabBcc  AAbBCc
            aaBbCc  AaBBcc  AAbBcC
            aaBBcc  aabBCC  AAbbCC
            aAbbcC  aaBBcc  AaBBCc
            aAbbCc  aaBBCc  AaBBcC
            aAbBcc  aAbbcC  AaBbCC
      aabbcC  aABcc  aAbBcc  AabBCC  AABBcC
      aabbCc  Aabbcc  aABBcc  aABBCc  AABBCc
      aabBcc  AabbCc  AAbbcC  aABBcC  AAbBCC
      aaBbcc  AabBcc  AAbBcc  aAbBCC  AABbCC
      aAbbcc  AaBbcc  AAbBcc  aAbBCC  aABBCC
aabbcc  Aabbcc  AAbbcc  AABbcc  aaBBCC  AaBBCC  AABBCC
  500    520    540    560    580    600    620
```

In this case each single upper case allele adds only 20 kg to the base weight of 500 kg/plot. This is determined in the same way as for the one and two gene models, although it is considerably more involved. You should note, once more that the F_1 would have had a yield potential of 560 kg/plot, exactly the same as in the single and two gene cases.

Even with only three loci and two alleles at each locus, it should be obvious that we are coming closer to a shape resembling a standard normal distribution. The frequency of different genotypes possible when four, five and six loci are considered has 9, 11 and 12 phenotypic classes, respectively. It is fairly easy to visualize, therefore, that with only a modest number of loci with segregating alleles acting in a more or less equal additive way, truly continuous variation in a character would be quite closely approached. Quantitative inheritance deals with many loci and alleles, often too many to consider estimating, and therefore explains the ubiquity of the normal distribution. Just as the mean and variance can describe the normal distribution, many of the important elements of the inheritance of a character can be described and explained using progeny means and genetic variances.

Quantitative genetics models

The relationship, and importance of the normal distribution, to quantitative genetics is clear, however, the closeness of the relationship between observed progeny performances and theoretical distributions will be related to **the model** on which the relationship is based. For example, we assumed that all uppercase letter alleles were of equal additive value, which of course may not be true. It is important that an appropriate model of inheritance is applied; otherwise other derived statistics (i.e. heritabilities, see later in Chapter 6) which are potentially of great value in plant breeding will be biased, and are likely to be highly misleading.

Let us examine the basic model applied to quantitative situations and see how the model can be tested for its appropriateness to the situation or inheritance of specific characters.

Consider again the cross between two canola cultivars described above. The yield of the higher yielding parent (P_1) is 620 kg/plot, the yield of the lower yielding parent (P_2) is 500 kg/plot, and the yield of the F_1 is 560 kg/plot. Assume a model of additive genetic effects, where we have:

$$
\begin{array}{ccc}
\bar{P}_2 & \bar{F}_1 & \bar{P}_1 \\
500 & 560 & 620 \\
\end{array}
$$
$$\xleftarrow{\hspace{2cm}} m \xrightarrow{\hspace{2cm}}$$
$$\xleftarrow{} [a] \xrightarrow{} \quad \xleftarrow{} [a] \xrightarrow{}$$

where the difference between the performance of the parents is divided in half (i.e. $120/2 = 60$ kg $= [a]$), and indicates the additive effect. Note that $[a]$ carries no sign. m is called the mid-parent value and is midway in value between the performance of $\bar{P}_1$ and $\bar{P}_2$. Therefore:

$$[a] = (\bar{P}_1 - \bar{P}_2)/2$$

$$m = \bar{P}_2 + [a] \quad \text{or} \quad \bar{P}_1 - [a]$$

$$\bar{P}_1 = m + [a] \quad \text{and} \quad \bar{P}_2 = m - [a]$$

The term $[a]$ is used to indicate the summation of the additive effects over all loci involved, however many this may be. In the example shown we assumed a completely additive model of inheritance and the $\bar{F}_1$ performance was indeed equal to m. Therefore in the absence of dominance, the mid-parent value will equal the performance of the $\bar{F}_1$. Dominance will be detected in cases where the performance of the $\bar{F}_1$ is not equal to m.

Let us return again to the canola cross, and continue to assume the relationship between uppercase alleles adding to a base yield and lowercase alleles adding nothing. Previously, we did not consider dominant alleles and their effect on the distribution.

Assume a two loci and two alleles per locus model of inheritance for yield. Assume also that A is dominant to a, but B and b are additive. Therefore $Aa = AA$, so we have:

$$P_2 \times P_1$$
$$AABB \times aabb$$

One (or two) A alleles would add 60 kg/plot to the base weight.

Therefore AA adds 60 kg/plot, $Aa = AA$ (dominance) $= 60$ kg/plot, and B adds 30 kg/plot. The $\bar{F}_1 = AaBb = 590$ kg/plot. Now $[a] = (\bar{P}_1 - \bar{P}_2)/2 = 60$ kg, so $m = \bar{P}_2 + [a] = 560$, clearly the $\bar{F}_1$ is not equal to m, and we have a case of dominance.

$$\begin{array}{ccc} \bar{P}_2 & \bar{F}_1 & \bar{P}_1 \\ 500 & 590 & 620 \end{array}$$
$$\longleftarrow \quad m \quad \longrightarrow$$
$$\longleftarrow [a] \longrightarrow \longleftarrow [a] \longrightarrow$$

When the F_2 population is examined we see that the basic bell-shape curve has now been skewed to the right (below), as a greater frequency of progeny have higher yield due to the effect of the dominant A allele. The average (mean) performance of the F_2 is now 575 kg/plot.

				AABb	
			AAbb	AaBb	
			Aabb	aABb	aABB
		aabB	aAbb	AaBB	AaBB
	aabb	aaBb	aaBB	aAbB	AABB
	500	530	560	590	620

↑

Expand this idea on to a three loci, two allele example as before and we have 64 possible genotypes with 7 possible phenotypes. Assume, that A is dominant to a, but that B, b, C and c are all additive, and uppercase alleles add 20 kg/plot to the base yield of 500 kg/plot. We now have, the $\bar{F}_1 = AaBbCc = 580$ kg/plot (as $Aa = AA = 20 + 20 = 40$ kg/plot). Again we see that the $\bar{F}_1$ performance is higher than the mid-parent value (m) indicating dominance.

As with the two loci case, the distribution of phenotypes with three genes is similarly skewed to the right. In this instance, the average (mean) performance of the F_2 generation would be 570 kg/plot.

					AABBcc	
					AABbcC	
					AABbCc	
			aabBCC	AAbBCc		
			aaBBcC	AAbBcC		
			aaBBCc	AAbbCC		
			aaBbCC	aAbBcC		
			AAbbcC	aAbbCC	AABBcC	
			AAbbCc	aAbbCC	AABBCc	
			AAbBcc	AabbCC	AAbBCC	
		aabbCC	AABbcc	AabBcC	AABbCC	
		aabBcC	aAbbcC	AabBCc	AaBBcC	
		aabBCc	aAbbCc	AaBbcC	AaBBcC	
		aabbcC	aAbBcc	AaBbCc	AaBBCC	
		aaBbCc	aAbBcc	AaBBcc	AabBCC	
	aabbcC	aaBBcc	AabbcC	aAbbcC	aABBCc	
	aabbCc	AAbbcc	AabbCc	aAbBcC	aABBcC	AABBCC
	aabBcc	aAbbcc	AabBcc	aABBcc	aABbCC	aABBCC
aabbcc	aaBbcc	Aabcc	AaBbcc	aaBBCC	aAbBCC	AaBBCC
500	520	540	560	580	600	620

↑

The keen observer will have noted two points:

- The F_1 performance was proportionally higher in the two gene case compared to the three gene case, because a higher proportion of alleles in the three gene case were showing non-dominance. In Figure 5.3, the degree of skewness of a six loci two allele system are shown for no dominance, one dominant loci, three dominant loci and five dominant loci.

- There was a relationship between the parent performance and the performance of the F_1 and F_2. The mean performance of the F_2 is midway between m, the mid-parent value (560 kg/plot) and the mean of the F_1 family (580 kg/plot). Not surprisingly the mean of the F_3 family would be 565 kg/plot, half-way between the mid-parent and the F_2 values. The mean

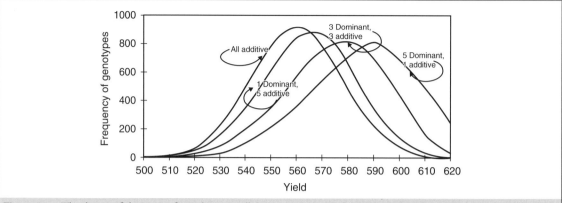

Figure 5.3 The degree of skewness of a six loci two-allele system is shown for no dominance, one dominant locus, three dominant loci and five dominant loci. Note that increasing the number of dominant loci results in greater skewness.

of the F_∞, of recombinant inbred lines would be equal to the mid-parental value, the same as if no dominance (or indeed linkage) existed. Therefore, if dominant effects are adversely effecting selection in plant breeding, then increased rounds of selfing can eliminate these effects.

Consider again the inheritance model that we have for additive effects:

$$\begin{array}{ccc} \bar{P}_2 & \bar{F}_1 & \bar{P}_1 \\ 500 & 560 & 620 \end{array}$$
$$\xleftarrow{\hspace{1.3cm}} m \xrightarrow{\hspace{1.3cm}}$$
$$\xleftarrow{} [a] \xrightarrow{} \quad \xleftarrow{} [a] \xrightarrow{}$$

As we have seen the $\bar{F}_1$ performance does not always coincide with the mid-parent value. In the two loci case we had:

$$\begin{array}{cccc} \bar{P}_2 & & \bar{F}_1 & \bar{P}_1 \\ 500 & & 560 \quad 580 & 620 \\ & & m \quad\quad [d] & \end{array}$$

So we now need to add a second parameter to the model [d] which represents the amount of dominance, and where:

$$[d] = \bar{F}_1 - m$$

and from this we have:

$$\bar{F}_1 = m + [d]$$

Unlike [a] which is always positive, [d] can be either positive or negative (i.e. the $\bar{F}_1$ has higher or lower performance than m, respectively).

Using these three parameters we can now proceed to determine the expected performance of any generation. We have:

$$\text{High parent } (\bar{P}_1) = m + a;$$

$$\text{Low parent } (\bar{P}_2) = m - a;$$

$$\bar{F}_1 = m + d$$

Assume that the $\bar{P}_1, \bar{P}_2$ and $\bar{F}_1$ generations are grown so that m, [a] and [d] can be calculated from their means, then it can thus be seen why m, a and d are referred to as the *components of the generation means*.

$$\begin{array}{ccc} \bar{P}_2 & \bar{F}_1 & \bar{P}_1 \\ m - [a] & m + [a] & m + [a] \end{array}$$
$$\xleftarrow{\hspace{2.5cm}} m \xrightarrow{\hspace{2.5cm}}$$
$$\xleftarrow{\hspace{0.8cm}} [a] \xrightarrow{\hspace{0.8cm}} \quad \xleftarrow{\hspace{0.8cm}} [a] \xrightarrow{\hspace{0.8cm}}$$

In addition to $\bar{P}_1, \bar{P}_2$ and $\bar{F}_1$ three further generations are commonly considered:

- The F_2 generation (i.e. F_1 selfed)
- The *back-cross generation* $F_1 \times P_1$, which is termed the B_1 generation
- The *back-cross generation* $F_1 \times P_2$, which is termed the B_2 generation

These three generations are typified by genetic segregation. It is therefore necessary to derive the proportions of the different genotypes, and their relative contributions to the means, in the various generations.

The Aa gene segregates with gamete formation in the F_1 thus giving, in the F_2 generation, the genotypes AA, Aa and aa in the ratio $1\ AA : 2\ Aa : 1\ aa$ or, as proportions, $\frac{1}{4} AA, \frac{1}{2} Aa$ and $\frac{1}{4} aa$. The mean expression of the AA plants is $m + a$. But, since only one quarter of the F_2 generation is of the AA genotype, it contributes $1/4(m + a)$ to the F_2 generation mean. Similarly, since the mean expression of the aa plants is $m - a$ and they make up $\frac{1}{4}$ of the F_2 generation, aa plants contribute $\frac{1}{4}(m - a)$ to the F_2 generation mean. Finally, since half the F_2 generation has the genotype Aa which has a mean expression of $m + d$, the heterozygotes contribute $\frac{1}{2}(m + d)$ to the F_2 generation mean. The term $[a]$ is a summation of additive effects over all loci therefore, for simplicity we assume (without proof):

$$\bar{F}_2 = \frac{1}{4}\bar{P}_1 + \frac{1}{2}\bar{F}_1 + \frac{1}{4}\bar{P}_2$$

$$= \frac{1}{4}(m + [a]) + \frac{1}{2}(m + [d]) + \frac{1}{4}(m - [a])$$

$$= \frac{1}{4}m + \frac{1}{4}[a] + \frac{1}{2}m + \frac{1}{2}[d] + \frac{1}{4}m - \frac{1}{4}[a]$$

$$= m + \frac{1}{2}[d]$$

Considering a single gene model again, we have the B_1 generation is composed of $\frac{1}{2} AA$ and $\frac{1}{2} Aa$. Again assuming without proof that $[a]$ is an accumulation of all additive effects and $[d]$ an accumulation of all dominant effects. We have:

$$\bar{B}_1 = \frac{1}{2}\bar{P}_1 + \frac{1}{2}\bar{F}_1$$

$$= \frac{1}{2}(m + [a]) + \frac{1}{2}(m + [d])$$

$$= m + \frac{1}{2}[a] + \frac{1}{2}[d]$$

Similarly the B_2 generation would be:

$$\bar{B}_2 = m - \frac{1}{2}[a] + \frac{1}{2}[d]$$

TESTING THE MODELS

Earlier, a model which we shall now call the **additive–dominance model**, was put forward that purports to explain the inheritance of quantitative (continuously varying) characters exclusively in terms of the additive and dominance properties of the single gene differences which underlie it. It is now necessary to consider whether models based on only additive and dominance genetic differences are adequate. From a plant breeding standpoint it is important to know if the inheritance of a character under selection is controlled by an additive–dominance model because many assumptions, notably the response to selection, are based on heritability, and estimating heritability is usually assumed that this model is suitable. Testing the additive–dominance model in quantitative genetics should be regarded as directly comparable to testing for the absence of linkage or epistasis with qualitative inheritance. However, with genes showing qualitative differences the most common testing method to compare frequencies of genotypes or phenotypes is a χ^2 test. In quantitative genetics genotypes (or phenotypes) do not fall into distinct classes and hence frequency χ^2 tests are not appropriate. So we need an equivalent but appropriate test.

First you should recall that the components m, $[a]$ and $[d]$ are derived from the generation means of the $\bar{P}_1, \bar{P}_2$ and $\bar{F}_1$ generations:

$$\bar{P}_1 = m + [a] : \bar{P}_2 = m - [a] : \bar{F}_1 = m + [d]$$

and from these similar equations can be formulated for the generation means of the $\bar{F}_2, \bar{B}_1$ and $\bar{B}_2$ generations:

$$\bar{F}_2 = m + \frac{1}{2}[d] : \bar{B}_1 = m + \frac{1}{2}[a] + \frac{1}{2}[d] :$$

$$\bar{B}_2 = m - \frac{1}{2}[a] + \frac{1}{2}[d]$$

These six equations lead to various *predictive* relationships between the means of different combinations of the generations. For example, if the additive–dominance model is correct then it can be predicted that:

$$2\bar{B}_1 - \bar{F}_1 - \bar{P}_1 = 0$$

This relationship is easily proved by the substitution of the appropriate combinations of m, $[a]$ and $[d]$ for the three generation means:

$$2\left(m + \frac{1}{2}[a] + \frac{1}{2}[d]\right) - (m + [d]) - (m + [a]) = 0$$

$$(2m + [a] + [d]) - (m + [d]) - (m + [a]) = 0$$

$$2m + [a] + [d] - m - [d] - m - [a] = 0$$

It can be seen that all the terms on the left-hand side of the last equation cancel out. This test is called the *A-scaling test*.

Another relationship is:

$$2\bar{B}_2 - \bar{F}_1 - \bar{P}_2 = 0$$

and:

$$2\left(m - \frac{1}{2}[a] + \frac{1}{2}[d]\right) - (m + [d]) - (m - [a]) = 0$$

$$(2m - [a] + [d]) - (m + [d]) - (m - [a]) = 0$$

$$2m - [a] + [d] - m - [d] - m + [a] = 0$$

Once again, all the terms on the left-hand side cancel out. This test is called the *B-scaling test*.

A final relationship, known as the *C-scaling test*, is:

$$4\bar{F}_2 - 2\bar{F}_1 - \bar{P}_1 - \bar{P}_2 = 0$$

These relationships are based on the predicted means of the various generations. Would you expect this relationship to hold if we substituted the means that we had actually measured? In other words would you expect the above equations to equal zero exactly? In fact, it would be quite surprising if, for example, the sum of the means of the $\bar{P}_2$ and $\bar{F}_1$ generations exactly equalled twice the mean of the $\bar{B}_2$ generation. While $\bar{P}_2 + \bar{F}_1$ might be approximately equal to $2\bar{B}_2$, error variation would give rise to random variation in all three means resulting in some overall discrepancy.

Thus, thinking now in terms of measured generation means:

$$2\bar{B}_1 - \bar{F}_1 - \bar{P}_1 = A$$

$$2\bar{B}_2 - \bar{F}_1 - \bar{P}_2 = B$$

$$4\bar{F}_2 - 2\bar{F}_1 - \bar{P}_1 - \bar{P}_2 = C$$

where, A, B and C are all expected to equal zero. If they do equal zero, or at least are not too far from it, then there is no reason to suspect that the additive–dominant model is inadequate as an explanation of the inheritance of the continuously varying character. On the other hand, if they do deviate markedly from zero, then there is reason to doubt the adequacy of the model as an explanation of the inheritance of the character in question. This is a classic instance of the need for objective statistical tests to decide whether A, B or C differ significantly from zero, or whether any discrepancies observed could be due to chance. If the discrepancies could reasonably

be attributed to chance, then the model can be provisionally accepted as an adequate description of reality. If this explanation is too unlikely, then the hypotheses (that A, B and C all equal zero within the bounds of sampling error) must be rejected along with the additive–dominant model on which they are based. The statistical tests are called the A-scaling test, B-scaling test and C-scaling test (and there are many others). However, the A-scaling test will be used here to represent the principles involved in them all.

The A-scaling test

The basis of the A-scaling test is that the value of A is compared to the value predicted on the assumption that an additive–dominant model is adequate (i.e. A = 0). The question is then asked:

> If the hypothesis is true (i.e. A = 0), what is the probability that any difference between observation (A) and prediction (zero) could be due to chance?

Conventionally, if the probability that the difference was due to chance is less than 0.05 (i.e. 5%, or 1 in 20) then the null hypothesis (in this case, that A *is equal to* 0) is rejected and the alternative hypothesis (that A $\neq$ 0) is accepted. In accepting the alternative hypothesis, it is also accepted that $2\bar{B}_1 - \bar{F}_1 - \bar{P}_1$ is not equal to zero, and that an additive–dominant model is inadequate in this particular instance.

In comparing the actual value of A with its predicted value (zero), what factors must be taken into account? Clearly the magnitude of the discrepancy (i.e. the actual value of A itself, $2\bar{B}_1 - \bar{P}_1 - \bar{F}_1$) must be considered. Another important factor is the *variability in A* from one experiment to another. If A varied enormously from one experiment to the next, then the mean value of A would have to be relatively large for it to be significantly different from zero. On the other hand, if the value of A were relatively constant from experiment to experiment, then even quite a small value of A could be accepted as significantly different from zero.

Finally, values based on relatively few plants are likely to be less convincing than values based on the measurement of many plants. Thus sample size is also highly relevant.

All of this perhaps sounds like a pretty tall order. In fact, statisticians have provided us with a method of relating the difference between the actual and predicted values of A to the variability in A from experiment to experiment, and also a statistical table in which the probability of obtaining such a difference by chance, given the number of plants measured, can be looked up. The equation is:

$$t = \frac{\text{Actual value of A - predicted value of A}}{\text{standard error of A}}$$

$$t = \frac{A - 0}{se(A)}$$

$$t = \frac{A}{se(A)}$$

In fact, as I am sure that you have all noticed, the A-scaling test is just a particular application of Student's t test, which you may have come across elsewhere.

In order to calculate t, A has to be divided by its *standard error* (se). It is known that $A = 2\bar{B}_1 - \bar{F}_1 - \bar{P}_1$, where $\bar{B}_1$, $\bar{F}_1$ and $\bar{P}_1$, are the measured (not the **predicted**) means of the B_1, F_1 and P_1 generations. But what is the standard error of A? A standard error is, like a standard deviation, the square root of a variance as shown earlier. In fact, the standard error of A is the square root of the variance of the mean of A (σ_A^2), and is represented as σ_A. Therefore:

$$\sigma_A^2 = 4\sigma_{\bar{B}_1}^2 + \sigma_{\bar{F}_1}^2 + \sigma_{\bar{P}_1}^2$$

where $\sigma_{\bar{B}_1}^2$ is the variance of the mean of the B_1 generation, $\sigma_{\bar{F}_1}^2$ is the variance of the mean of the F_1 generation and $\sigma_{\bar{P}_1}^2$ is the variance of the mean of the P_1 generation.

It is essential to realize that the variance of the mean of a generation (i.e. $\sigma_{\bar{B}_1}^2$) is not the same as the variance between plants in that generation. In principle, the former is calculated by growing adequate numbers of plants representing the generation in several different experiments, calculating a generation mean for each experiment and then calculating the variance of these different means (effectively treating them as raw data). A variance of the mean so determined is less than the variance between all the plants grown in all the experiments. Fortunately, it is not necessary to perform several different experiments as described. Statisticians have demonstrated that a satisfactory estimate of the variance of the mean is obtained by dividing the variance derived from a single sample of plants by the number of plants measured that contribute to the estimate of the

mean. That is:

$$\sigma_{\bar{B}_1}^2 = \frac{\text{variance of } n\ B_1\ \text{plants}}{n}$$

As an example if this consider that the height of 50 individual plants (i.e. $n = 50$) of a pure-line barley cultivar was recorded and that the average plant height of all plants measured was calculated to be 100 cm, with a variance of 40 cm^2. The variance and the standard error of the mean of this sample of plant heights would be:

$$\text{Variance of mean} = \frac{\text{variance of 50 plants}}{50}$$

$$= \frac{25}{50} = 0.80\ \text{cm}^2$$

And the standard error $= \sigma = \sqrt{(0.80)} = 0.89$ cm

Therefore the mean is 100 cm $\pm$ 0.89 cm.

So, given a set of individual measurements, you should be able to calculate the mean, the variance and the standard deviation of the population of which the data you are given can be assumed to be an unbiased sample. You should also be able to calculate both the variance of the mean and thus its standard error.

Now consider a simple example where two homozygous barley cultivars (P_1 and P_2) are cross pollinated and a sample of F_1 seed is backcrossed to the higher yielding parent (P_1) to produce B_1 seed. Now if 11 P_1, 11 F_1 and 11 B_1 seeds were planted in a properly randomized experiment and the height of each plant recorded and the following means and standard errors are determined.

$$\bar{P}_1 = 109.1 \pm 9.1 : \bar{F}_1 = 105.5 \pm 8.6 :$$
$$\bar{B}_1 = 108.3 \pm 10.0$$

Now the variance of each family would be:

$$\sigma_{\bar{P}_1}^2 = (9.1)^2 : \sigma_{\bar{F}_1}^2 = (8.6)^2 : \sigma_{\bar{B}_1}^2 = (10.0)^2$$

Therefore it follows that the variance of A (σ_A^2) would be:

$$\sigma_A^2 = 4\sigma_{\bar{B}_1}^2 + \sigma_{\bar{F}_1}^2 + \sigma_{\bar{P}_1}^2$$
$$= 4(10.0)^2 + (8.6)^2 + (9.1)^2 = 200$$

and the standard error of A (σ_A) is given by:

$$\sigma_A = \sqrt{(\sigma_A^2)} = \sqrt{(200)} = 14.1$$

Now to consider the mean value of A, this is given by:

$$A = 2\bar{B}_1 - \bar{F}_1 - \bar{P}_1$$
$$= 2(107.3) - 105.5 - 109.1 = 2.0$$

Finally:

$$t = A/\sigma_A = 2.0/14.1 = 0.142.$$

So, a value of t has been calculated for these data. This is based on both the deviation of A from its expected value of zero (i.e. 2.0) and the variability found in the P_1, F_1 and B_1 plants measured, all the variability being summarized in the standard error of A (i.e. 14.1). The question now is the deviation statistically significant?

In order to decide this, it is necessary to account for the number of plants measured in each generation on which the values of A and σ_A are based. In fact, it is not the number of plants as such that is used but the relevant numbers of degrees of freedom, where the degrees of freedom of A = the degrees of freedom of $\bar{B}_1$ + the degrees of freedom of $\bar{F}_1$ + the degrees of freedom of P_1.

Degrees of freedom have been previously mentioned in connection with the χ^2 test. Generally, the number of degrees of freedom associated with a generation is one fewer than the number of plants representing that generation. Thus, if 11 plants of each of generations B_1, F_1 and P_1 were measured then the degrees of freedom of A = $(11-1) + (11-1) + (11-1) = 30$ df.

It is necessary to look up the value of t (i.e. 0.142) for 30 degrees of freedom in a table of probabilities for t. As the t value we obtained is smaller in magnitude compared to the table value with 30 degrees of freedom so there is no reason to reject the additive-dominant model in this instance, and so it is provisionally accepted as an adequate explanation of the inheritance of the character in question.

Joint scaling test

The procedure described above can be repeated in a similar way to derive a test for the B-scaling test or the C-scaling test. Indeed, sets of such scaling tests can be devised to cover any combination of types of family that may be available.

As an alternative, however, to testing the various expected relationships one at a time, a procedure was proposed by a researcher called Cavalli, which is known as the *joint scaling test*. This test effectively combines the whole set of scaling tests into one and thus offers a more general, more convenient, more adaptable and more informative approach.

The joint scaling test consists of estimating the model's parameters, m, $[a]$ and $[d]$ from the means of all types of families available, followed by a comparison of these observed means with their expected values derived from the estimates of the three parameters. This makes it clear at once that at least three types of family are necessary if the parameters of the model are to be estimated. However, with only three types of family available no test can be made of the goodness of fit of the model since in such a case a perfect fit **must** be obtained between the observed means and their expectations from the estimates of the three parameters. So to provide such a test, at least four types of family must be raised.

The procedure for the joint scaling test is illustrated by considering the example given by Mather and Jinks (*Introduction to Biometrical Genetics*). The data they presented have been truncated for simplicity and so differences due to rounding errors may occur. Their example consists of a cross between two pure-breeding varieties of rough tobacco (*Nicotiana rustica*). The means and variances of the means for plant height of the parental, F_1, F_2 and first back-cross families (B_1 and B_2) derived from this cross are shown in Table 5.7.

Also shown in this table is the number of plants that were evaluated from each generation. Family size was deliberately varied with the kind of family. It was set at as low as 20 for the genetically uniform parents and in excess of 100 for the F_2 and back-crosses, to compensate for the greater variation expected in these segregating families. All plants were individually randomized at the time of sowing so that the variation within families reflects all the non-heritable sources of variation to which the experiment is exposed. With this design the estimate of variance of a family mean (V_x), valid for use in the joint scaling test, is obtained in the usual way by dividing the variance within the family by the number of individuals in that family. Reference to these variances shows that the greater family size of the segregating generations has more than compensated for their greater expected variability in that the variances of their family means are smaller than those of their non-segregating families.

Table 5.7 Means and variances of the means for plant height of two parental lines ($\bar{P}_1$ and $\bar{P}_2$), the $\bar{F}_1$, $\bar{F}_2$ progeny, and the first back-cross families ($\bar{B}_1$ and $\bar{B}_2$) derived from crossing $\bar{P}_1$ to $\bar{P}_2$.

	Number of plants	V_x	Weight $1/V_x$	Model m	Model $[a]$	Model $[d]$	Observed
$\bar{P}_1$	20	1.033	0.968	1	1	0	116.30
$\bar{P}_2$	20	1.452	0.669	1	-1	0	98.45
$\bar{F}_1$	60	0.970	1.031	1	0	1	117.67
$\bar{F}_2$	160	0.492	2.034	1	0	1/2	111.78
$\bar{B}_1$	120	0.489	2.046	1	1/2	1/2	116.00
$\bar{B}_2$	120	0.613	1.630	1	−1/2	1/2	109.16

Table 5.8 Coefficients of m, $[a]$ and $[d]$ in the parents ($\bar{P}_1$ and $\bar{P}_2$), the $\bar{F}_1$, $\bar{F}_2$, generation and both back-cross generations ($\bar{B}_1$ and $\bar{B}_2$) and the observed plant height of each family.

Generation	Model m	Model $[a]$	Model $[d]$	Observed
$\bar{P}_1$	1	1	0	116.30
$\bar{P}_2$	1	−1	0	98.45
$\bar{F}_1$	1	0	1	117.67
$\bar{F}_2$	1	0	1/2	111.78
$\bar{B}_1$	1	1/2	1/2	116.00
$\bar{B}_2$	1	−1/2	1/2	109.16

Six equations are available for estimating m, $[a]$ and $[d]$ and these are obtained by equating the observed family means to their expectations as given above. The coefficients of m, $[a]$ and $[d]$ in the six equations are listed with the collected data. These coefficients are shown in Table 5.8.

There are three more equations than there are parameters to be estimated (m, $[a]$ and $[d]$), therefore a least square technique can be used. The six generation means to which we are fitting the m, $[a]$ and $[d]$ model are not known with equal precision; for example, the variance of the mean (V_{P_2}) of P_2 is almost three times that of the B_1. The best estimates will be obtained, therefore, if generation means and are weighted in relation to how accurate the estimates are. The appropriate weights in this instance are the reciprocals of the variances of the means. For the first entry in the data (above) P_1, the

weight is given by $1/1.0334 = 0.9677$ and so on for the other families.

The six equations and their weights may be combined to give three equations whose solution will lead to weighted least squares estimates of m, $[a]$ and $[d]$, as follows. In order to obtain the first of these three equations each of the six equations is multiplied through by the coefficient of m that it contains, and by its weight, and the six are then summed. When we weight each line of the array by m (which is always equal to 1) the sum total we have:

m	$[a]$	$[d]$		Observed
+0.9677	+0.9677	0	=	112.541
+0.6688	−0.6688	0	=	65.848
+1.0310	0	+1.0310	=	121.327
+2.034	0	+1.0171	=	227.376
+2.0458	+1.0229	+1.0229	=	237.316
+1.6300	−0.8150	+0.8150	=	177.931
$\sum$ 8.3775	+0.5067	+3.8860	=	942.340

The second and third equations are found in the same way using the coefficient of $[a]$ for the second and of $[d]$ for the third along with, the weights ($1/V_x$) as multipliers.

To illustrate, the next line is found in the same way by multiplying each of the lines by the coefficients of

[a] (i.e. +1, −1, 0, 0, +1/2, −1/2), and then summing columns thus:

m	[a]	[d]		Observed
+0.9677	+0.9677	0	=	112.541
−0.6688	+0.6688	0	=	−65.848
0	0	0	=	0
0	0	0	=	0
+1.0229	+0.5115	+0.5115	=	118.658
−0.8150	+0.4075	−0.4075	=	−58.965
$\sum$ 0.5067	+2.5555	+0.1040	=	76.385

Finally the third line is obtained by multiplying through by the coefficients of [d] (i.e. 0, 0, 1, 1/2, 1/2), and then summing the columns thus:

m	[a]	[d]		Observed
0	0	0	=	0
0	0	0	=	0
+1.0310	0	+1.0310	=	121.327
+1.0171	0	+0.5085	=	113.688
+1.0229	+0.5114	+0.5114	=	118.658
+0.8150	−0.4075	+0.4075	=	99.965
$\sum$ 3.8860	+0.1040	+2.4585	=	442.639

We then have three simultaneous equations, known as normal equations, which may be solved in a variety of ways to yield estimates of m, $[a]$ and $[d]$. A general approach to the solution is by way of matrix inversion. The three equations are rewritten in the form:

$$\begin{bmatrix} 8.3775 & 0.5067 & 3.8860 \\ 0.5067 & 2.5555 & 0.1040 \\ 3.8860 & 0.1040 & 2.4585 \end{bmatrix} \times \begin{bmatrix} m \\ [a] \\ [d] \end{bmatrix} = \begin{bmatrix} 942.340 \\ 76.385 \\ 442.638 \end{bmatrix}$$
$$\mathbf{J} \qquad\qquad \times\ \mathbf{M}\ =\ \mathbf{S}$$

where $\mathbf{J}$ is known as the information matrix, $\mathbf{M}$ is the estimate of the parameters and $\mathbf{S}$ is the matrix of the scores.

The solution then takes the general form $\mathbf{M} = \mathbf{J}^{-1} \times \mathbf{S}$ where $\mathbf{J}^{-1}$ is the inverse of the information matrix and is itself a variance–covariance matrix.

The inversion may be achieved by any one of a number of standard procedures, for our example, inversion leads to the following solution:

$$\begin{bmatrix} m \\ [a] \\ [d] \end{bmatrix} = \begin{bmatrix} 0.4568 & -0.0613 & -0.7194 \\ -0.0613 & 0.4002 & 0.0800 \\ -0.7194 & 0.0800 & 1.5405 \end{bmatrix} \times \begin{bmatrix} 942.339 \\ 76.385 \\ 442.639 \end{bmatrix}$$
$$\mathbf{M}\ =\qquad\qquad \mathbf{J}^{-1} \qquad\qquad\quad \times\quad \mathbf{S}$$

The estimate of m is then:

$$m = (0.4568 \times 942.339) - (0.0613 \times 76.385)$$
$$- (0.7194 \times 442.639)$$
$$= 107.322$$

The standard error (s.e.) of m is $\sqrt{(0.4568)} = \pm 0.6759$.

In a similar way:

$$[a] = 8.1997 \pm 0.6326$$
$$\text{and} \quad [d] = 10.0587 \pm 1.2412$$

All are highly significantly different from zero when looked up in a table of normal deviates.

The adequacy of the additive-dominance model may now be tested by predicting the six family means from the estimates of m, $[a]$ and $[d]$.

For example:

$$\bar{B}_2 = m - \frac{1}{2}[a] + \frac{1}{2}[d]$$

on the basis of this model and for the estimates obtained it has as the expected value:

$$107.3220 - [1/2 \times 8.1997] + [1/2 \times 10.0597]$$
$$= 108.2515$$

This expectation along with those for the other five families is listed in Table 5.9.

The agreement with the observed values appears to be very close and in no case is the deviation more than 0.83% of the observed value. The goodness of fit of this model can be tested statistically by a χ^2. Since the data comprise six observed means, and three parameters

Table 5.9 Observed plant heights from both parents ($\bar{P}_1$ and $\bar{P}_2$), the $\bar{F}_1$, $\bar{F}_2$, generations and both backcross generations ($\bar{B}_1$ and $\bar{B}_2$) along with the expected plant height from the joint scaling test parameters, and the difference between the observed and expected plant height.

Family	Observed	Expected	Obs–Exp
$\bar{P}_1$	116.300	115.522	+0.778
$\bar{P}_2$	98.450	99.122	+0.672
$\bar{F}_1$	117.675	117.381	+0.294
$\bar{F}_2$	111.778	112.351	−0.573
$\bar{B}_1$	116.000	116.451	−0.451
$\bar{B}_2$	109.161	108.252	−0.090

have been estimated (i.e. m, $[a]$ and $[d]$), thus the χ^2 value has $6 - 3 = 3$ degrees of freedom.

The contribution made to the χ^2 by $\bar{P}_1$, for example, is the difference between Observed and Expected divided by the variance (or in our case we can multiply by one over the variance that is the weight. So, for example $[116.300 - 115.522]^2 \times 0.968 = 0.5862$. Summing the six such contributions, one from each of the six types of family, gives a χ^2 of 3.411 for 3 degrees of freedom, which has a probability of between 0.40 and 0.30. The model must therefore be regarded as adequate (i.e. there is no evidence of anything beyond additive and dominance effects).

The individual scaling tests, A, B and C, referred to earlier can, of course, also be used to test the model. Thus with the present data

$$A = 2\bar{B}_1 - \bar{P}_1 - \bar{F}_1$$
$$= [2 \times 116.000] - 116.300 - 117.675$$
$$= -1.975$$
$$\sigma_A^2 = 4\sigma_{B_1}^2 + \sigma_{P_1}^2 + \sigma_{F_1}^2$$
$$= [4 \times 0.4888] + 1.0334 + 0.9699$$
$$= 3.959$$

leading to $\sigma_A = \sqrt{(3.959)} = 1.990$

Thus $A = -1.98 \pm 1.99$ which, when compared in Student t statistical tables does not differ significantly from the value 0 expected.

The joint scaling test gives exactly the same answer as the A-, B- or C-scaling tests. However, the joint scaling test does more than test the adequacy of the additive–dominance model. It also provides the '*best*' estimates of all the parameters required (and their standard errors) to account for differences among family means when the model is adequate. If you try to estimate m, $[a]$ and $[d]$ with the procedure shown earlier you will find values of $m = 107.365$, $[a] = 8.925$ and $[d] = 10.310$. These estimates do not differ markedly from those estimated (107.322, 8.1997, 10.0587, respectively) from the joint scaling test. However, the difference may in some cases be of importance. The joint scaling test can also be readily extended to more complex situations.

To conclude, in this example the best estimates show that the additive and dominance components are of the same order of magnitude and since $[d]$ is significantly positive, alleles that increase final height must be dominant more often than alleles that decrease it.

What could be wrong with the model

In some instances the results from generation testing will conclude that an additive–dominant model of inheritance does not adequately account for the data. There are many possible explanations of this, and here only three, in order of increasing genetic complexity, will be mentioned briefly.

- **Abnormal chromosomal behaviour.** In the case of single gene or multiple gene cases, the predicted means of the $\bar{F}_1$, $\bar{B}_1$ and $\bar{P}_1$ generations were derived in terms of m, $[a]$ and $[d]$. Throughout, the assumption was made that a heterozygote contributes equal proportions of its various gametes to the gene pool following genetic segregation. If this assumption is unwarranted. then the frequencies of the different genotypes, and therefore their contributions to the generation means in terms of m, $[a]$ and $[d]$, would not be as predicted. This could result from the elimination of deleterious alleles through gametic selection.
- **Cytoplasmic inheritance.** If a character is determined, or affected, by non-nuclear genes, then the character will depend on which of the genotypes was maternal. Under such circumstances, inheritance cannot be explained fully in terms of m, $[a]$ and $[d]$.
- **Non-allelic interactions or epistasis.** We have already mentioned the phenomenon of genetic dominance.

Effectively, dominance-recessiveness are allelic interactions whereby the phenotypic expression of a character does not depend solely on the additive effects of the different alleles at the same locus, and the mean expression of the heterozygote is not the same as the mid-parent value. As we note earlier, a similar phenomenon can occur between alleles at different loci, and this is known as **non-allelic interaction** or **epistasis**. An example of epistasis and how it might occur was presented in the qualitative genetics section. A further example might be:

$$AABB = 24; \quad AAbb = 12; \quad aaBB = 12; \quad aabb = 8$$

In the presence of *BB*, the difference between the *AA* and *aa* genotypes is $24 - 12 = 12$ units. However, in the presence of *bb*, the difference between *AA* and *aa* is $12 - 8 = 4$ units. Of course, another way of looking at the matter might be to say that the difference between *BB* and *bb* is $24 - 12 = 12$ units in the presence of *AA*, but $12 - 8 = 4$ units in the presence of *aa*. Either way, it can be seen that there is **interaction** between the alleles at different loci and that an additive-dominant model of inheritance cannot be adequate. In fact, it is possible to add epistasis (usually symbolized by *aa*, for interaction between loci which are homozygous, *ad* for those between loci where one is heterozygous and one homozygous and *dd* for loci which are heterozygous).

In general it is actually quite straightforward to take into account other genetic phenomena by inclusion of appropriate parameters in the basic additive–dominant model of inheritance and thus increasingly account for more complex genetic inheritance.

Although you should be aware of the existence of these complications, they will not be taken any further in this book. Moreover, it is often found that, for most metrical characters of interest to plant breeders the additive–dominant model is adequate – if it fails we are then aware that the situation is more complex and act accordingly. Also, since what is of primary practical interest is the ratio of the additive genetic variance in a generation to the variance attributable to all causes (environmental, additive, dominant and all other genetic phenomena), it is often unnecessary to itemize them individually.

QUANTITATIVE TRIAL LOCI

The concept of linkage, between different loci located on the same chromosome, was introduced in the qualitative genetics section. Quantitatively inherited characters are controlled by alleles at multiple loci. Yield, for example, is a highly complex character which is related to a multitude of other characters like, seedling germination and emergence, flowering times, partition, photosynthesis efficiency, nitrogen uptake efficiency, etc., plus a susceptibility or resistance to a wide range of stresses including diseases and pests. Even if a single gene were to be responsible for all the individual factors that are involved in yield potential (which they are not), then it is easy to see that there will be hundreds or even thousands of genes which influence yield. Given that the number of chromosomes in crop species is small ($2n = 2x = 34$ in sunflower, $2n = 2x = 18$ in lettuce, $2n = 4x = 38$ in rapeseed, $2n = 2x = 20$ in maize, $2n = 6x = 42$ in wheat, $2n = 2x = 14$ in barley, $2n = 2x = 24$ in rice, $2n = 2x = 22$ in bean, and $2n = 4x = 48$ in potato), then linkage will always be a major factor in the inheritance of quantitatively inherited traits. So this, as with other quantitative effects, adds another level of complexity. In general, the complexity of the genetics has meant that many questions remain unanswered. Some questions that might be asked are:

- Whether all loci have equal effect on the quantitative trial expression or whether there are some of the loci that have major effects while others have minor effects
- Whether the multiple loci are distributed evenly throughout the genome, or whether they are clustered on specific chromosomes, or in specific regions of the genome ('hot spots')

The concept of quantitative trait loci (QTL) was first raised by Sax in 1923. Sax reported examining yield on a segregating F_2 progeny from a cross between two homozygous common bean (*Phaseolus vulgaris*) lines. One parent was homozygous for coloured seed while the other had white seed. A single gene at the P-locus determined seed colour, with *PP* alleles for coloured seed and *pp* for white seed. On inspection of seed weights, Sax found that *PP* lines produced seeds with an average weight of 30.7 g/100 seeds, heterozygotes (*Pp*) produced seed with 28.3 g/100 seeds, while *pp* lines had smallest seed weights (26.4 g/100 seeds). From this Sax introduced the concept that the quantitative loci determining seed weight were linked to the single gene locus for seed colour.

The potential of expanding this concept in plant breeding attracted the attention of many researchers after Sax's work was published. However, few advantages were achieved because plant breeders were forced to work with mainly morphologically visible single gene traits and major-gene mutants. These were not the most suitable for investigating QTL's because:

- They were relatively few in number
- Were usually recessive, and their expression masked in the phenotype by dominant alleles
- Often had deleterious effects (or pleitropic effects) on the quantitative trait of interest (i.e. albinism, dwarfism, etc.)

These defects have been corrected by the introduction of molecular markers, which tend to be numerous, do not affect the plant phenotype, and are often co-dominant allowing the heterozygotes to be differentiated from the homozygotes parental types.

In plant breeding, QTL's have greatest potential in marker assisted selection for quantitatively inherited traits which have low heritability or that are difficult or expensive to screen or evaluate.

The process involved in QTL's will be illustrated using a simple, simulated, example where two homozygous parents are hybridized to produce F_1 plants. One parent was homozygous *AABBCC* at the A-, B- and C-bands, respectively, while the other parent was homozygous *aabbcc*. It should be noted that in this example, *A* is not dominant to *a*, etc.

Thirty two homozygous lines were derived from the F_1 family using double-haploidy techniques (see Chapter 8). These lines were grown in a four replicate field trial to determine yield of each line. In addition, the lines were polymorphic for three loci that appeared to be located on the same chromosome. The molecular marker banding at the three molecular markers (identified simply as A, B and C-bands, *AA*, *BB*, and *CC*, respectively) along with the yield of each line is shown in Table 5.10. We use doubled haploids in this example for simplicity as there will be no heterozygotes in the population. This makes some of the calculations simpler as dominance effects can be ignored. However, the principle is the same and can be carried out using any segregating population resulting from a two-parent cross.

Mapping of the three qualitative loci is done according to the method described earlier, and the map is as follows:

A-------31.3 mμ-------B----18 mμ----C

Table 5.10 Yield of 32 double haploid lines, and genotype of each line at the A-, B-, and C-bands.

Line	A-band	B-band	C-band	Yield	Line	A-band	B-band	C-band	Yield
1	*AA*	*BB*	*CC*	107.80	17	*aa*	*BB*	*cc*	112.41
2	*AA*	*BB*	*CC*	113.57	18	*aa*	*bb*	*cc*	104.93
3	*AA*	*BB*	*cc*	111.68	19	*aa*	*bb*	*cc*	104.62
4	*aa*	*bb*	*CC*	101.09	20	*AA*	*BB*	*CC*	114.68
5	*aa*	*bb*	*cc*	91.29	21	*AA*	*BB*	*CC*	110.79
6	*aa*	*bb*	*cc*	112.24	22	*AA*	*bb*	*cc*	101.47
7	*aa*	*bb*	*cc*	97.17	23	*AA*	*BB*	*cc*	116.61
8	*aa*	*bb*	*cc*	95.75	24	*aa*	*bb*	*CC*	101.95
9	*aa*	*BB*	*CC*	113.52	25	*aa*	*bb*	*cc*	106.33
10	*aa*	*BB*	*CC*	119.27	26	*aa*	*bb*	*cc*	95.42
11	*AA*	*bb*	*cc*	98.40	27	*AA*	*BB*	*CC*	121.85
12	*AA*	*BB*	*CC*	106.82	28	*AA*	*BB*	*CC*	111.94
13	*AA*	*BB*	*CC*	117.61	29	*AA*	*bb*	*cc*	105.45
14	*AA*	*BB*	*CC*	112.88	30	*AA*	*bb*	*cc*	99.15
15	*AA*	*bb*	*CC*	101.58	31	*aa*	*BB*	*CC*	116.49
16	*aa*	*BB*	*CC*	119.27	32	*aa*	*bb*	*cc*	100.21

Table 5.11 Degrees of freedom and mean squares from the analysis of variance of seed yield on 32 double haploid lines grown in a three replicate randomized complete block design.

Source	df	MSq
Between lines	31	1241.8 ***
Replicate blocks	3	321.1 ns
Replicate error	93	401.4

which, when converted to cMs is:

A-------49.5 cM-------B----22.5 cM----C

The first stage in QTL analysis is to determine if there are indeed significant differences between the progeny lines. This is done by carrying out a simple analysis of variance. In our example, there were indeed significant differences between these lines (see Table 5.11).

Where there are significant differences in yield detected between the parental lines, can this difference in yield potential be explained by association between yield and the single marker bands?

Assume, for simplicity here, that genotypes with A-bands have genotype *AA*, and those without have genotype *aa*, and similarly for the B- and C-bands. Average yield of each single band genotype can be calculated by adding the yield of lines carrying the same bands at each locus and dividing by the number of individual lines in that class. For example, the average of all lines, which have the *AA* bands, is 109.52, while those that have the *aa* bands is 105.23. Similarly, yield of the *BB* band types is 114.31 compared to 100.44 for *bb*, and *CC* types is 112.05 compared to *cc* types which are 102.70. From this, there appears a pattern that lines carrying the *BB* band rather than the *bb* band have the largest yield advantage. Similarly, lines carrying the *CC* band over the *cc* band also have an advantage (albeit smaller than with the B-band). *AA* and *aa* lines differ only slightly. To apply significance to these differences requires partition of the sum of squares for differences between lines is partitioned into:

- Sum of squares due to the difference between the two genotypes at each band position, BG–SS (i.e. between *AA* types and *aa* types; *BB* and *bb* types; and *CC* and *cc* types)

- Sum of squares due to the variation between lines within each genotype at each band, WG–SS

In this simple example, there are 16 lines that are *AA* and 16 with *aa*. Similarly there are 16 lines that are *BB*, *bb*, *CC* and *cc*. Therefore it is completely balanced. In this instance the partition of the lines' sum of squares is by a simple orthogonal contrast. In actual experiments, the number of individuals in each class is likely to vary, and the BG–SS partition is completed by:

$$BG\text{–}SS = \left\{ \frac{(\bar{x}_{11} \cdot n_1)^2}{n_1} + \frac{(\bar{x}_{22} \cdot n_2)^2}{n_2} \right. \\ \left. - \frac{[(\bar{x}_{11} \cdot n_1) + (\bar{x}_{22} \cdot n_2)]^2}{(n_1 + n_2)} \right\} \Big/ \text{number of reps}$$

Where $\bar{x}_{11}$ is the mean of lines with the 11 genotype, and n_1 is the number of lines with the 11 genotype. In this example, for the *AA* and *aa* genotypes we would have:

$$BG\text{–}SS = \left\{ \frac{(\bar{x}_{AA} \cdot n_A)^2}{n_A} + \frac{(\bar{x}_{aa} \cdot n_a)^2}{n_a} \right. \\ \left. - \frac{[(\bar{x}_{AA} \cdot n_A) + (\bar{x}_{aa} \cdot n_a)]^2]}{(n_A + n_a)} \right\} \Big/ 4$$

$$= \left\{ \frac{(109.52 \times 16)^2}{16} + \frac{(105.23 \times 16)^2}{16} \right. \\ \left. - \frac{[(109.52 \times 16) + (105.23 \times 16)]^2}{(16 + 16)} \right\} \Big/ 4$$

$$= 2351$$

The sum of squares for variation within genotypes (WG–SS) is obtained by subtracting the variation between types (above) from the total sum of squares between lines:

$$WG - SS = SS \text{ lines} - BG - SS$$

In the case of the *AA* and *aa* bands we have:

$$WG - SS = 38\,498 - 2351 = 36\,147$$

The degree of freedom for the between genotype sum of squares is one, while the degrees of freedom for the within genotype sum of squares is the total number of lines minus two, (in our example $32 - 2 = 30$).

Completing this operation for the other two bands, we have the mean squares from three analyses of variance (see Table 5.12).

Table 5.12 Degrees of freedom and mean squares from the analyses of variance of seed yield between and within progeny that are polymorphic at the *AA* : *aa*, *BB* : *bb* and *CC* : *cc* loci.

Source	df	*AA–aa* locus	*BB–bb* locus	*CC–cc* locus
Between genotypes	1	2351 ns	24 618***	11 211***
Within genotypes	30	36 147***	463 ns	913*
Replicate error	93	459	459	459

The within genotypes effect is tested against the replicate error, while the between genotype effect is tested against the within genotype mean square.

Clearly, there is a significant relationship between seed yield and alleles at the B-bands. Similarly, some relationship exists between the C-band and yield, although the variability with genotypes *CC* and *cc* are highly significant, hence weakening the QTL relationship. There is no relationship between seed yield and bands at the A band.

From the above analysis of variance, our best guess to the position of the QTL would be between the B and C bands, and nearer to the B than the C. Determination of the position of the QTL on the chromosome can be done using a number of statistical techniques. The simplest technique involves regression, and will be illustrated here.

Now the difference in yield between genotypes at each band (δ_i) is an indication of the linkage between the QTL and the single band position. In this example we have:

$$\delta_{A-a} = 2.145; \quad \delta_{B-b} = 6.935; \quad \delta_{C-c} = 4.475$$

Given a simple additive–dominance model of inheritance, we find that lines with *BB*, plus the QTL + will have expectation of $m + a$ and this genotype will occur in the population with frequency $\frac{1}{2}R$, where R is the recombination frequency between the B-band and the QTL. The *BB* lines without the QTL (QTL-) will be $m - a$, and will occur in the population with frequency $\frac{1}{2}(1 - R)$, where R is the recombination frequency between the B-band and the QTL. Similarly, for *bb* we have, *bb*/QTL+ = $m + a$, frequency = $\frac{1}{2}R$, *bb*/QTL- = $m - a$, frequency = $\frac{1}{2}(1 - R)$. The difference between the *BB* and *bb* genotypes (δ_{B-b}) is therefore equal to $a(1 - 2R)$.

There is therefore a linear relationship between δ_i's and $(1 - 2R)$,

$$\delta_i = a(1 - 2R)$$

where the regression slope is an estimate of a.

The value of a and the accuracy of fit of regression is dependant on R, the recombination frequency between the three single band position and the QTL. We know the map distance between A–B, and B–C. Therefore all that is now required is to substitute in recombination frequencies to find the recombination frequency which has the least departure from regression in a regression analysis of variance. It is usual to start with the assumption that the QTL is located at the A-band position and complete a regression analysis. Then assume that the QTL is 2 cM from A, towards B, and carry out another analysis. Repeat this operation until it is assumed that the QTL is located at the C-locus. Thereafter determine which of the regression analyses has the best regression fit (with least departure from regression term) and the QTL will be located at that map location. From this the recombination frequencies between the various single band position and the QTL can be calculated to determine the usefulness of the linkage with the QTL and hence the usefulness in practice.

To avoid duplication, the speculative map distances from each single band position and the QTL in our example is shown from around the map location with minimum departure from regression are shown:

	A-locus	B-locus	C-locus	Residual sum of squares
δ_i	2.145	6.935	4.475	
R_i's	0.300	0.013	0.193	1932
	0.310	0.003	0.183	802
	0.3198	0.01478	0.17148	0
	0.330	0.017	0.163	339

From this resulting map including the QTL would be:

A --- 30.5 mμ --- B -- 1.5 mμ -- QTL -- 17.1 mμ --- C which, when converted to cM's is:

A --- 47.1 cM --- B -- 1.5 cM -- QTL --- 21.0 cM --- C

In this simple case the QTL and the B-bands are tightly linked and therefore selection based on the B-band would be highly effective in selecting for the QTL, and hence high seed yield. Actual examples in plant breeding, however, are rarely this close. The close proximity of the QTL to the B-band is also a reflection of the high recombination frequencies (low linkage) between the three bands in this simple example. If the recombination frequencies between A–B and B–C were halved (i.e. 15.7% and 9.0%) the QTL would have a recombination of 3% with the B-band position. Similarly, this example looked at only three bands on a single chromosome, in real situations, many chromosomes will be involved and more loci examined on each chromosome. However, the underlying theory is the same.

THINK QUESTIONS

(1) Explain what may cause a departure from a $9:3:3:1$ expected frequency of phenotypes. Describe an appropriate statistical test to **prove** your hypothesis.

(2) Past researchers have shown that an additive–dominance model can explain the inheritance of plant height in spring canola (*Brassica napus* L.). Below are shown family means, the standard errors (s.e.) of the mean and the number of plants that these data are estimated from $\bar{P}_1$, $\bar{P}_2$, $\bar{F}_1$ and $\bar{F}_2$ families.

Family	Mean	s.e. of mean	Number of plants
$\bar{P}_1$	40.2	0.142	31
$\bar{P}_2$	19.3	0.151	31
$\bar{F}_1$	35.4	0.099	31
$\bar{F}_2$	28.7	0.462	31

Using an appropriate statistical test determine if the interpretation of past researchers hold for these phenotypes.

If the additive–dominance model does not explain the variation found for plant height, give three reasons for what might be the cause?

(3) Explain (using *A* as a dominant allele and *a* as a recessive allele) the difference between genotypes that are nulliplex, simplex, duplex, triplex and quadruplex for a single dominant gene in a tetraploid. Assuming no complications such as double reduction, what would be the expected ratio of nulliplex, simplex, duplex, triplex and quadruplex resulting from a cross between two auto-tetraploid lines that are duplex for a single gene (i.e. *AAaa* × *AAaa*).

(4) Explain the meaning of a '*test cross*' as applied to testing linkage disequilibrium in qualitative genetics.

Two homozygous barley genotypes are chosen for a linkage study. One of the parents had a long awn and was short in stature. The other parent had a short awn and was tall in stature. Long awn is controlled by a single dominant gene (*AA* = Long awn is dominant over *aa*) and plant height is controlled by a single dominant gene (*TT* = Tall is completely dominant over *tt*). The two lines are crossed (i.e. *ttAA* × *TTaa*) and the resulting F_1 is test crossed to a homozygous genotype with short awn and short stature (*ttaa*). 4000 plants from the test cross are grown and the following phenotypic frequencies were observed:

Tall and Long awn(*TA*) = 396

Short and Long awn(*tA*) = 1610

Tall and Short awn(*Ta*) = 1590

Short and Short awn(*ta*) = 404

Determine the recombination percentage.

(5) Two homozygous squash plants were hybridized and an F_1 family produced. One parent has long, green fruit (*LLGG*), and the other had round, yellow fruit (*llgg*). 1600 F_2 progeny were examined from selfing the F_1s and the following numbers of phenotypes observed:

LLGG	*L-gg*	*llG-*	*llgg*
891	312	0	397

Complete an appropriate analysis to explain and interpret this segregation pattern.

(6) Assuming an additive–dominance model of inheritance is adequate to explain the variation in the cross between two parents (P_1 and P_2), both are diploid and homozygous genotypes. In terms of m, $[a]$ and $[d]$, what would be the expected performance of each parent ($\bar{P}_1$ and $\bar{P}_2$) and the $\bar{F}_1$ family? Also in terms of m, $[a]$ and $[d]$, what would be the expected values of the F_2 population and the two possible backcrosses (B_1 and B_2)? Using simple algebra, derive the expected values you have listed for the $\bar{F}_2$ family, above, in terms of the frequency of segregating plant types in that generation.

In a single cross between two homozygous barley cultivars (P_1 and P_2) the yield of $\bar{P}_1$ was 2642 kg/ha, the yield of $\bar{P}_2$ was found to be 1290 kg/ha and the $\bar{F}_1$ family showed a yield of 2308 kg/ha. Given that the additive–dominance model is adequate to explain genetic variation for yield in barley, what would the expected yield of the $\bar{F}_2$, $\bar{B}_1$ and $\bar{B}_2$ families be?

(7) Cultivated potatoes (*Solanum tuberosum* spp. *tuberosum*) are all auto-tetraploids and therefore diploid segregation of qualitative traits is not appropriate. A new single, completely dominant, gene has been identified which confers resistance to Colorado beetle. Given two parental lines, each with a single copy of this new gene, are inter-crossed and hybrid seed produced. What proportion of the hybrid family would you expect to be resistant to potato Colorado beetle and what would be the expected frequency of nulliplex, simplex, duplex, triplex and quadruplex genotypes in the progeny?

The aim of multiplex breeding is to develop parental lines that have either three or four copies of a single dominant gene of interest. Starting with two simplex parents design a scheme to develop either a triplex or quadruplex parental line. Include in the scheme the frequency of genotypes obtained at each stage and also include any test crosses that may be necessary.

(8) The yield data were collected on 32 recombinant inbred lines of barley from a cross between two homozygous parents (*AABBCC* × *aabbcc*), that

were grown in a replicated yield trial. Also shown is the genotype of each inbred lines at the A-, B- and C-locus (note that capital letters do not infer dominant alleles, and lowercase letter do not refer to recessive alleles). The analysis of variance of yield is presented below. Determine whether these three loci are located on the same chromosome and whether any of the loci are linked to QTLs for yield.

Line	A-locus	B-locus	C-locus	Yield
1	AA	BB	CC	30.1
2	AA	BB	CC	20.8
3	AA	BB	cc	27.4
4	aa	bb	cc	13.7
5	aa	bb	cc	24.1
6	AA	BB	cc	21.8
7	aa	bb	cc	16.9
8	aa	bb	cc	13.5
9	AA	BB	CC	32.6
10	AA	BB	CC	33.0
11	aa	bb	cc	23.8
12	AA	BB	CC	26.7
13	AA	BB	CC	26.4
14	AA	bb	cc	11.1
15	AA	BB	CC	24.8
16	aa	BB	CC	31.9
17	aa	BB	cc	21.7
18	aa	bb	cc	22.1
19	aa	bb	cc	15.0
20	AA	BB	CC	25.4
21	AA	BB	CC	29.3
22	AA	bb	cc	20.8
23	AA	BB	cc	21.0
24	aa	bb	CC	19.7
25	AA	bb	CC	14.5
26	aa	bb	cc	17.0
27	AA	BB	CC	29.7
28	AA	BB	CC	32.0
29	aa	bb	cc	16.4
30	aa	bb	cc	14.0
31	aa	BB	CC	31.7
32	aa	bb	cc	16.9

(9) A new barley disease (yellow stripe rust, YSR) has recently been identified in the Palouse region. A single, completely dominant gene has been found which confers complete and durable resistance to this disease. A successful barley variety should be dwarf (controlled by a single recessive gene) and be shatter resistant (controlled by a single recessive gene). Using the notation Y to indicate a single YSR resistance gene and y to be a susceptible recessive gene, T to be a single dominant tall gene and t to be a single dwarf gene, S to be a shatter susceptible gene and s to be a recessive resistant gene. Two crosses were examined:

$(1) = YYTT \times yytt$ and $(2) = YYSS \times yyss$

The F_1 family from each cross was test crossed to the recessive parent and a number of progeny screened. From cross (1) the following results were found:

> YSR resistant and tall = 3240
>
> YSR resistant and short = 3100
>
> YSR susceptible and tall = 3400
>
> YSR susceptible and short = 3260

(a) Given that the F_1 family was selfed to produce an F_2 and this in turn was selfed to produce F_3 families (without selection) what would be the expected frequency of genotypes and phenotypes be in the F_3 family?

(b) How many F_3 genotypes would need to be evaluated to be 99% sure of having at least one plant which is homozygous YSR resistant and dwarf?

From cross (2) the following results were obtained from the test cross (res. = resistant and susc. = susceptible):

> YSR res. and shatter susc. = 5000
>
> YSR res. and shatter res. = 1290
>
> YSR susc. and shatter susc. = 1310
>
> YSR susc. and shatter res. = 5400

(c) Given that the F_1 family was selfed to produce an F_2 (without selection) what would be the expected frequency of genotypes and phenotypes in the F_2 family?

(10) Both parents (P_1 and P_2), the F_1 and F_2 from a cross between two homozygous wheat lines were grown in a replicated field trial and the yield of a number of individual plants was recorded. From these data the total yield of all plants measured (Σx), sum of squares (Σx^2) and the number of plants assessed (n) are shown below:

	$\bar{P}_1$	$\bar{P}_2$	$\bar{F}_1$	$\bar{F}_2$
Σx	324	166	305	591
Σx^2	10 000	30 000	9 100	21 678
n	11	11	11	21

From the data above, determine the mean and variance of each family and then test if the additive–dominance model is adequate to explain the inheritance for yield in this cross. Suggest three reasons why the additive–dominance model of inheritance might not be adequate to explain the inheritance of a character.

(11) Three different F_2 populations were grown out from a cross between a dwarf susceptible parent and a tall resistant parent ($ttrr \times TTRR$) and plants assessed for dwarfism and disease resistance. The number of plants which were observed in the four possible phenotype classes is shown below from three different trials:

	$T\_R\_$	T_rr	$ttR\_$	$ttrr$
(1)	8960	3002	3050	988
(2)	9041	3009	0	3950
(3)	0	12 001	2997	1002

It is known that these two genes are located on different chromosomes. Explain what could have caused the segregation pattern in each example (i.e. (1), (2) and (3)). Using a suitable chi-square test, determine if the segregation pattern for example (1) in part (a), are correct.

(12) One parent (P_1), the F_1 and B_1 families were grown in a replicated yield trial. From each population, plant height (mm) was recorded on 10 plants 35 days after planting. The mean and

standard error of the mean (s.e.) are shown below:

	Mean	s.e.
$\bar{P}_1$	244	2.00
$\bar{F}_1$	217	2.41
$\bar{B}_1$	232	3.64

Using an appropriate scaling test, determine if an additive–dominance model is adequate to describe the observed variation in this character.

When breeding field beans, it is very difficult to obtain large numbers of F_1 seed. Therefore scaling tests which include the F_1 or either backcross (B_1 of B_2) is prohibitive. Design an appropriate scaling test using the F_2, F_3 families and both parents (P_1 and P_2) and using m, $[a]$ and $[d]$, show

that the expectation of your test should be equal to zero given an additive–dominance model.

(13) A cross is made between two homozygous barley parents. One parent is tall and resistant to mildew (i.e. *TTRR*) and the other parent is short and susceptible to mildew (i.e. *ttrr*). Height and mildew resistance are both qualitatively inherited with tall being dominant to short, and resistant dominant to susceptible. If the F_1 family from this cross were self pollinated how many F_2 plants would you need to grow to ensure a 99% certainty of having **at least** one plant which was resistant to mildew and short. Consider now the same cross (*TTRR* × *ttrr*) and in a breeding programme 6400 F_2 plants were evaluated. At harvest only the short mildew resistant plants were selected and grown as F_3 head rows (i.e. only these selected types were evaluated at the F_3 stage). Determine the expected *number of genotypes and phenotypes* you would have when you harvest the F_3 rows.

6

Predictions

INTRODUCTION

Plant breeders strive to make a wide range of predictions which allow them to act in the most effective way in creating genetic variation and selecting desirable genotypes.

There are a variety of questions that might be posed, but ones that a breeder might sensibly ask would include:

- What expression of what traits would be most successful when a cultivar is released?
- What will the cultivar need to display to meet agronomic requirements so as to fit ideally into the most effective management systems?
- How stable will be the expression of the important traits (e.g. yield) over a range of environments (especially locations and years)?
- Which parents will give the best progeny for further breeding or for commercial exploitation?
- Which traits will respond most significantly to the selection imposed?
- What type and level of selection will give the optimal response in the traits of interest?

The first question is clearly difficult to answer and is one that faces breeders all the time. The second question is partly a matter of selection conditions, partly a matter of judging what is required and partly luck! The third is one on which a considerable amount of work has been carried out and is, of course, one of genotype by environment interaction. It is an important aspect of breeding but despite its importance we intend to only spend a limited amount of time on it.

However, questions such as the last three (above) need to be answered by a combination of knowledge

of genetics, experimental design and statistics – the better our knowledge the more accurate should be our predictions!

Let us start by considering the stability of expression over environments and its genetic determination.

Genotype × environment interactions

The performance (in other words its phenotype) of a genotype will differ in different environments. This is straightforward and so if we give less fertilizer we generally get less yield, the more we space the plants of many species the more vegetative growth they make. This can clearly be handled in our selection programme by selecting under the conditions we think are most appropriate and if necessary in several different ones. But what is more complicated is that not all genotypes respond to the same extent or necessarily in the same way to differences in the numerous environmental variables.

So some genotypes are more drought tolerant, some more disease susceptible, some can withstand higher levels of salt than counterparts, and so on. We cannot grow all the possible genotypes we are interested in under all the possible environmental conditions. We might note that some environmental variables are of a 'macro' nature and fairly obvious, but there are numerous possible differences in the environment that are experienced by individual plants – 'micro' ones.

So winter or spring sowings give a clear set of environmental differences that we might take specifically into account, just as different latitudes, temperature, semi-arid and tropical climates might be important considerations in an international breeding programme. But differences in water availability at one end of test area compared to the other, the row spacing produced by one piece of farm equipment compared with another, are but a few of the possible subtle differences which might be very important to individuals or plots or fields or farms.

There are a number of ways that plant breeders try to take genotype × environment interactions (G × E) into account in breeding but it does mean that there a is need to carry out trials over a range of environments which might be simply different sites or over years/seasons or running trials with defined differences such as water levels etc. Genotype × environment at its simplest can be examined by looking at the variance (or standard error by taking the square root) of the phenotypes over the range of environments and selecting for the lowest variance as being the most stable – remembering that we also need a good mean expression! Or a much more sophisticated approach is to use the mean of all the material grown to provide a biological measure of the environment and compare (usually by regression – slope of the line) the individual lines, families, clones, etc. against this.

It is not appropriate to go into greater details here about G × E or the various possibilities to take it into account. Needless to say many breeding programmes effectively ignore G × E in an explicit way but take it into account to a modest extent by trialling the more advanced material at different sites, by the fact that selection is carried out over a number of years, etc. A more detailed examination of analyses of multiple year and multiple location trials is presented in Chapter 7.

Let us now consider prediction protocols associated with answering which parents will give the best progeny and which traits will respond most significantly to the selection we impose?

Genetically based predictions

Plant breeders use all the genetic information (qualitative and quantitative) in just the same way as we use the information from Mendelian Genetics – in other words to predict the properties of generations or families that have not actually been observed. So from an analysis of the observed variation, firstly determine how much of the variation is due to environmental effects and how much due to genetic effects. Often it is desirable to go further and separate the genetic into additive, dominance, and other genetic effects, and to determine

which direction is dominance acting in, and to what degree.

Let us consider one particular use of the information, how can we predict the response to selection? Before you start a selection programme you would obviously like to know what sort of response you might expect for any given input, which traits are worth targeting, which populations or crosses are best to use. Is it a worthwhile venture? These questions involve many aspects of the biology of the crop, its handling in agriculture, the availability of other methods to affect crops, for example, chemicals such as pesticides, herbicides etc. But one of the main components that will determine the outcome is the amount and type of variation that is present. For example, in the extreme case of no genetic variation the breeder is wasting time trying to select superior genotypes!

If genetic variation is present, but small compared with that due to the environment, then progress can be made but only very slowly, unless very large numbers are handled – in other words much of the time the breeder will be selecting phenotypes which are 'superior' but as this is mostly due to the environment it will not give a reliable indication of a '*superior*' genotype. If on the other hand the phenotype is a good reflection of the genotype, that is most of the variation is genetic, then progress will be quick, since when the breeder selects a good phenotype and uses it as a parent it will pass on the superior attributes (*via* its genes) to its offspring.

One obvious question to ask is can we estimate how much of the variation we observe, in for example the F_2 we were looking at, could be ascribed to genetic differences of segregating genes and how much to environmental causes? We therefore need to measure the proportion of the total observed variation, that is genetic variation, and such a measure is called *heritability*.

HERITABILITY

For a modern plant breeder (or indeed a farmer with no knowledge of genetics) to make progress in an organized programme of selective breeding, two conditions are a must:

- There must be some phenotypic variation within the crop. This would normally be expected, even if it were due entirely to the effects of a variable environment

- At least some of this phenotypic variation must have a genetic basis

This relates to the concept of heritability, the proportion of the phenotypic variance that is genetic in origin. This proportion is called the heritability and this section is concerned with the ways of estimating heritability.

Values of heritability (h^2) can range from zero to one. If h^2 is relatively high (e.g. close to 1) there is potential for a breeding programme to alter the mean expression of the character in future generations. On the other hand, if h^2 is close to zero, there will be little scope for advancement and there would probably be little point in trying to improve this character in a plant breeding programme.

There are three main ways of estimating heritability.

- Carrying out particular genetic crosses so that the resulting data can be partitioned into their genetic and environmental components
- Based on the direct measurement of the degree of resemblance between offspring and one, or both, of their parents. This is achieved by regression of the former onto the latter in the absence of selection
- Measuring the response of a population to given levels of selection (this will not be discussed until we cover selection later)

The essential background theory of heritability was presented in the previous quantitative genetics section.

Heritability is a ratio of genetic variance divided by total phenotypic variance. In a simple additive–dominance model of quantitative inheritance the total genetic variance will contain **dominance genetic variance** (denoted by V_D) and **additive genetic variance** (denoted by V_A). Dominance genetic variance is variation caused by heterozygotes in the population, while additive genetic variance is variation between homozygotes in the segregating population.

Broad-sense heritability (h_b^2) is the total genetic variance divided by the total phenotypic variance. The total genetic variance in an additive–dominance model is simply $V_A + V_D$. The total phenotypic variance is obtained by summing the genetic variance plus the environmental variance.

The degree of heterozygosity within segregating populations will be related to the number of selfing generations. Maximum heterozygosity will be found in the

F_1 family, and will be reduced, by half, in each subsequent selfed progeny. Similarly, the dominance genetic variance will be dependent on the degree of heterozygosity in the population and will differ between filial generations. A more useful form of heritability for plant breeders, therefore, is **narrow-sense heritability** (h_n^2), which is the ratio of *additive genetic variance* (V_A) to total phenotypic variance.

Why should lack of resemblance between parents and their offspring be attributable to dominance, but not additive, components? Well, dominance effects are a feature of particular genotypes; but, genotypes are 'made' and 'unmade' between generations as a result of genetic segregation during the production of gametes. Thus, the mean dominance effect in the offspring of a particular cross can be different from that of the parents, even when there is no selection. On the other hand, when selection is applied, there may be no change or even change in the 'wrong' direction. This is not true of additive genetic effects. The additive genetic component must remain more or less constant from one generation to the next in the absence of selection. While, if differential selection is applied, the change between generations must be in the direction corresponding to the favoured alleles. In addition, additive genetic variance is constant between filial generations and so narrow-sense heritability of recombinant inbred lines can be estimated from early-generation segregating families.

In the first filial generation, after hybridization between two homozygous parents (F_1), there is no genetic variance between progeny and all variation observed between F_1 plants will be entirely environmental. The first generation for which there are both genetic and environmental components of phenotypic variance is the F_2. Partitioning of phenotypic variance and the calculation of the broad-sense (and ultimately narrow-sense) heritabilities will be confined to this generation.

Broad-sense heritability

The first step is to derive an equation for the genetic variance of the F_2 generation. The genetic variance of the F_2 generation (without proof) is:

$$\sigma_{\bar{F}_2}^2 = \frac{1}{2} V_A + \frac{1}{4} V_D$$

The total phenotypic variance of any generation is the sum of its genetic variance plus the environmental variance, E. So the total phenotypic variance of the F_2 generation can be written:

$$\sigma_{F_2}^2 = \frac{1}{2}V_A + \frac{1}{4}V_D + \sigma_E^2$$

In terms of V_A, V_D and σ_E^2 therefore the broad-sense heritability of the F_2 generation is:

$$h_b^2 = \frac{\text{genetic variance}}{\text{total phenotypic variance}}$$

$$= \frac{\frac{1}{2}V_A + \frac{1}{4}V_D}{\frac{1}{2}V_A + \frac{1}{4}V_D + \sigma_E^2}$$

In order to estimate the broad-sense heritability of the F_2 family (or indeed any other segregating family) all that is required is an estimate of total phenotypic variation, and an estimate of environmental variation. The former is obtained by measurements on plants within F_2 families, while the latter is estimated from measurements on families or plants that have a uniform genotype (i.e. homozygous parental lines or F_1 families where plants are genetically identical and any variation between plants is due to environment).

To illustrate this, consider a simple numerical example. A field experiment with an inbreeding crop species was conducted which included 20 plants from Parent 1, 20 plants from Parent 2 and 100 plants from the F_2 family derived from selfing the F_1 generation obtained by inter-crossing the two parents. These 140 plants were completely randomized within the experiment and at harvest the weight of seeds from each plant recorded. The variances in seed weight of the two parents were, $\sigma_{P_1}^2 = 16.8 \text{ g}^2$ and $\sigma_{P_2}^2 = 18.4 \text{ g}^2$. The phenotypic variance (which included both genetic and environmental variation) of the F_2 was $\sigma_{F_2}^2 = 56.9 \text{ kg}^2$. Total phenotypic variance of the F_2 generation is $\frac{1}{2}V_A + \frac{1}{4}V_D + \sigma_E^2$ and is estimated to be 56.9 k^2 (the variance of the F_2). It therefore follows that the broad-sense heritability, h_b^2 for these data is:

$$h_b^2 = \frac{56.9 - \sigma_E^2}{56.9}$$

The problem reduces to: what is the value of the environmental component of the phenotypic variance, σ_E^2? Since, by definition, both parents are completely homozygous inbreds, any variance displayed by either must be attributable exclusively to the environment. The best estimate of the value of σ_E^2 is therefore the mean phenotypic variance of these two generations. Thus:

$$\sigma_E^2 = (16.8 + 18.4)/2 = 17.6 \text{ kg}^2$$

and

$$h_b^2 = \frac{56.9 - 17.6}{56.9} = 0.691$$

In other words, 69.1% of the phenotypic variance of the F_2 generation is estimated to be genetic in origin.

The other generation in which the phenotypic variance is also entirely attributable to environmental effects is the F_1. If the phenotypic variances of all three of these genotypically invariate generations were available, the environmental component of the phenotypic variance of the F_2 generation could be estimated as follows:

$$\sigma_E = \frac{\sigma_{P_1}^2 + 2\sigma_{F_1}^2 + \sigma_{P_2}^2}{4}$$

Research workers often use more elaborate formulae, but this one will serve our purpose.

Narrow-sense heritability

Often it is of more interest, for reasons already noted, to know what proportion of the total phenotypic variation is traceable to additive genetic effects rather than total genetic effects. This ratio of additive genetic variance to total phenotypic variance is called narrow-sense heritability (denoted by h_n^2) and is calculated as:

$$h_n^2 = \frac{\text{additive genetic variance}}{\text{total phenotypic variance}}$$

Therefore, in terms of V_A, V_D and σ_E^2, what is the narrow-sense heritability of the F_2 generation? Since:

$$h_b^2 = \frac{\frac{1}{2}V_A + \frac{1}{4}V_D}{\frac{1}{2}V_A + \frac{1}{4}V_D + \sigma_E^2}$$

it is reasonable to suppose that:

$$h_n^2 = \frac{\frac{1}{2}V_A}{\frac{1}{2}V_A + \frac{1}{4}V_D + \sigma_E^2}$$

In order to estimate the narrow-sense heritability it is therefore necessary to partition the genetic variance into its two components (V_A and V_D). This is done by considering the phenotypic variance of the two back cross families ($\sigma^2_{\bar{B}_1}$ and $\sigma^2_{\bar{B}_2}$). Without proof the expected variances of $\sigma^2_{\bar{B}_1}$ and $\sigma^2_{\bar{B}_2}$ are:

$$\sigma^2_{\bar{B}_1} = \frac{1}{4}V_A + \frac{1}{4}V_D - \frac{1}{2}[\Sigma(a) \times \Sigma(d)] + \sigma^2_E$$

$$\sigma^2_{\bar{B}_2} = \frac{1}{4}V_A + \frac{1}{4}V_D + \frac{1}{2}[\Sigma(a) \times \Sigma(d)] + \sigma^2_E$$

The awkward expression $\frac{1}{2}[\Sigma(a) \times \Sigma(d)]$ disappears when the equations are added together. Therefore:

$$\sigma^2_{\bar{B}_1} + \sigma^2_{\bar{B}_2} = \frac{1}{2}V_A + \frac{1}{2}V_D + 2\sigma^2_E$$

As it is also known that:

$$\sigma^2_{\bar{F}_2} = \frac{1}{2}V_A + \frac{1}{4}V_D + \sigma^2_E$$

Provided that numerical values for $\sigma^2_{\bar{F}_2}, \sigma^2_{\bar{B}_1}, \sigma^2_{\bar{B}_2}$ and σ^2_E can be estimated, there is sufficient information to calculate both V_A and V_D, and hence the narrow-sense heritability.

To illustrate this consider the following example. A properly designed glasshouse experiment was carried out with pea. Progeny from the F_1, F_2 and both backcross families (B_1 and B_2) were arranged as single plants in a completely randomized block design and plant height recorded after flowering. The following variances were calculated from the recorded data.

$$\sigma^2_{\bar{F}_2} = 358\,cm^2; \quad \sigma^2_{\bar{B}_1} = 285\,cm^2;$$

$$\sigma^2_{\bar{B}_2} = 251\,cm^2; \quad \sigma^2_E = 155\,cm^2$$

now:

$$\sigma^2_{\bar{B}_1} + \sigma^2_{\bar{B}_2} - \sigma^2_{\bar{F}_2} = \left(\frac{1}{2}V_A + \frac{1}{2}V_D + 2\sigma^2_E\right)$$
$$- \left(\frac{1}{2}V_A + \frac{1}{2}V_D + \sigma^2_E\right)$$
$$= \frac{1}{4}V_D + \sigma^2_E$$

and

$$V_D = 4(\sigma^2_{\bar{B}_1} + \sigma^2_{\bar{B}_2} - \sigma^2_{\bar{F}_2} - \sigma^2_E)$$
$$= 4(285 + 251 - 358 - 155)$$
$$= 92\,cm^2$$

Rearranging the equation for $\sigma^2_{\bar{F}_2}$ (i.e. $\sigma^2_{\bar{F}_2} = \frac{1}{2}V_A + \frac{1}{4}V_D + \sigma^2_E$) we have:

$$V_A = 2\left(\sigma^2_{\bar{F}_2} - \frac{1}{4}V_D - \sigma^2_E\right)$$
$$= 2\left(358 - \left[\frac{1}{4} \times 92\right] - 155\right)$$
$$= 360\,cm^2$$

Therefore, the narrow-sense heritability for these data is:

$$h^2_n = \frac{\frac{1}{2}V_A}{\frac{1}{2}V_A + \frac{1}{4}V_D + \sigma^2_E}$$
$$= \frac{0.5 \times 360}{0.5 \times 360 + 0.25 \times 92 + 155} = 0.50$$

This can be derived more simply by:

$$h^2_n = \frac{\frac{1}{2}V_A}{\text{total phenotypic variation}}$$
$$= \frac{\frac{1}{2}V_A}{\sigma^2_{\bar{F}_2}}$$
$$= \frac{0.5 \times 360}{358} = 0.50$$

Thus, 50% of the phenotypic variation in this F_2 generation of pea is genetically additive in origin.

Heritability from offspring–parent regression

In this section we will consider one other method of estimating the narrow-sense heritability. The option of predicting the response to selection using heritabilities will be discussed in the selection section (Chapter 7). However, the phenomenon does suggest another approach to measuring the heritability of a character, namely comparison of the phenotypes of offspring with those of one or both of their parents. Close correspondence in the absence of selection implies that the heritability must be relatively high. On the other hand, if the phenotypes appear to vary independently of one another, this suggests that heritability must be low.

The foundations of this approach, which is termed *offspring–parent regression*, were laid in the nineteenth century by Charles Darwin's cousin Francis Galton in his study of the resemblance between fathers and sons. Therefore, the narrow-sense heritability of a metrical character can be estimated from the regression coefficient of offspring phenotypes on those of their parents.

In regression analysis, one variable is regarded as **independent**, while another is regarded as being potentially **dependent** on it. Not surprisingly, in offspring–parent regression, the phenotype of the parent(s) corresponds to the independent variable and that of the offspring the dependent variable.

The narrow-sense heritability of a character in the F_2 generation is:

$$h_n^2 = \frac{\frac{1}{2}V_A}{\frac{1}{2}V_A + \frac{1}{4}V_D + \sigma_E^2}$$

and, the regression coefficient from simple linear regression is:

$$\mathbf{b} = SP(x, y)/SS(x)$$

Therefore, there must be some relationship between h_n^2 and the regression coefficient (**b**). The regression relationship when the offspring expression is regressed onto the expression of **one** of the parents (provided without proof) is:

$$\mathbf{b} = \frac{\frac{1}{4}V_A}{\frac{1}{2}V_A + \frac{1}{4}V_D + \sigma_E^2}$$

and since:

$$h_n^2 = \frac{\frac{1}{2}V_A}{\frac{1}{2}V_A + \frac{1}{4}V_D + \sigma_E^2}$$

it follows that, for the regression of offspring phenotypes on the phenotypes of one of their parents:

$$h_n^2 = 2 \times \mathbf{b}$$

In short, to estimate the narrow-sense heritability (h_n^2), it is necessary to perform a regression analysis of the mean phenotype of the offspring of individual parents on the phenotype of those parents. The regression coefficient (**b**) is obtained by dividing the offspring–parent covariance by the variance of the parental generation. The narrow-sense heritability is then double the value of the regression coefficient.

When the expression of progeny are regressed onto the average performance of **both** parents (the mid-parental performance) then the regression coefficient is (given without proof):

$$\mathbf{b} = \frac{\frac{1}{4}V_A}{\frac{1}{2}(\frac{1}{2}V_A + \frac{1}{4}V_D + \sigma_E^2)}$$

$$= \frac{\frac{1}{2}V_A}{\frac{1}{2}V_A + \frac{1}{4}V_D + \sigma_E^2}$$

and since:

$$h_n^2 = \frac{\frac{1}{2}V_A}{\frac{1}{2}V_A + \frac{1}{4}V_D + \sigma_E^2}$$

it follows that, for the regression of offspring phenotypes on the mean phenotypes of both their parents:

$$h_n^2 = \mathbf{b}$$

In short, to estimate the narrow-sense heritability it is necessary to perform a regression analysis of the mean phenotypes of offspring on the mean phenotypes of both their parents. The regression coefficient (**b**), obtained by dividing the offspring-parent covariance by the variance of the parental generation, estimates the narrow-sense heritability directly.

Consider the following simple example. The data below are the phenotypes of parents and their offspring from a number of crosses in a frost tolerant winter rapeseed breeding programme for yield (kg/ha).

Female parent	Male parent	Mid-parent value	Offspring value
995	1016	1005.5	1006
1004	999	1001.5	1004
1009	996	1002.5	1008
1012	1014	1013.0	1010
1005	1014	1009.5	1013
1007	1004	1005.5	1007
1034	1014	1024.0	1024
1015	998	1006.5	1002
1017	1028	1022.5	1020
1003	1013	1008.0	1008

Then

Statistic	Female parent	Male parent	Mid-parent value
Regression slope (**b**)	0.476	0.468	0.813
se(**b**)	0.1632	0.1895	0.1269
$t_{8 \text{ df}}$	2.898	2.259	6.407

From the regression of offspring on one parent:

$$\text{male } h_n^2 = 2 \times \mathbf{b} = 2 \times 0.473 = 0.946$$

$$\text{female } h_n^2 = 2 \times \mathbf{b} = 2 \times 0.468 = 0.936$$

From the regression of offspring onto the average phenotype of both parents (mid-parent) we have:

$$h_n^2 = 0.813$$

You will notice that the heritability using only one parent is larger than that from both parents. It should be noted that the estimation based on both parents will be more accurate. Despite the difference, however, it is obvious that there is a high degree of additive genetic variance for this character.

Finally, always remember that a heritability estimate, no matter which method is used to obtain it, is only valid for that population, at that time, in that environment! Change the environment, carry out (or allow) selection to occur, add more genotypes, sample another population and the heritability will be different! This should be clear from the descriptions and methods of calculating heritability but you will find many examples in the literature where the basic limitations of the concept are forgotten.

DIALLEL CROSSING DESIGNS

It has been over 130 years now since the publication by Louis de Vilmorin that became known as Vilmorin's

isolation principle or progeny test. He proposed that the only means to determine the value of an individual plant (or genotype) was to grow and evaluate its progeny. Ever since, of course, the progeny test has become well established and is frequently used by plant breeders to determine the genetic potential of parental lines. The diallel cross is simply a more sophisticated application to Vilmorin's progeny test.

The term *diallel cross* has been attributed to a Danish geneticist (J. Schmidt) who first used the design in animal breeding. The term and design came to plant breeding and began to be used by plant scientists in the mid 1950s.

The diallel cross was then described as *all possible crosses amongst a group of parent lines*. With n parents there would be n^2 families. The n^2 families or progeny are called a *complete diallel cross*. If the reciprocal crosses are not made, making $n[n-1]/2$ families, the result is called a *half diallel*. A *modified diallel* is one in which all possible cross combinations are included but the parental selfs (diagonal elements) are excluded. This type of diallel will include $n^2 - n$ families. In a *partial diallel* fewer than the $n[n-1]/2$ cross combinations are completed. However, the crosses that are included are arranged in such a way that valid statistical analysis and interpretation can be carried out.

Initially, only inbred homozygous lines were used as parents in diallel crossing designs. Techniques that allow for parents to be non-inbred genotypes (i.e. heterozygous) are now available.

According to some critics of the designs 'the diallel mating design has been used and abused more extensively than any other ...'. Whether this statement is true or otherwise, there is little doubt that if the theory of diallel analysis is adhered to and if interpretation can be carried out in a logical manner, then the use of diallel crossing designs can be of great benefit to plant breeders in aiding understanding of qualitative inheritance and providing invaluable information regarding the genetic potential of parental lines in cross combinations. The limitation of the design arises in terms of the sample of parental genotypes that can be handled, which is always somewhat restricted.

It will not be possible to cover the whole spectrum of information or even indeed the types of diallel crossing schemes that are available or to investigate the interpretation of many examples within the space available. There are therefore several approaches to the analysis and interpretation of diallel cross data although only two will be covered briefly in this section:

- Analysis of general and specific combining ability. These methods are often referred to as **Griffing analyses**, after B. Griffing who published his, now famous paper 'Concept of general and specific combining ability in relation to diallel crossing systems' in 1956.
- Analysis of array variances and covariances, often referred to as **Hayman and Jinks analyses**, after B.I. Hayman and J.L. Jinks' paper of 1953, 'The analysis of diallel crosses'.

Griffing's Analysis

Griffing proposed a diallel analysis technique for determining general combining ability and specific combining ability of a number of parental lines in cross combination based on statistical concepts. Griffing analyses have been used by many plant breeders and researchers over the past 40 years and in many cases with good success. Much of the success found in applying Griffing analyses is the apparent ease of interpretation of results compared to other analyses available. Parents used in diallel crosses can be homozygous or heterozygous, for simplicity diallel types are described here in terms of homozygous (inbred) parents. Four types of design analyses are available:

- **Method 1**. The full diallel where p parents are crossed in all possible cross combinations (including reciprocals). Therefore with p parents the design will consist of p^2 families ($p^2 - p$ segregating populations or F_1's and p inbred parents).
- **Method 2**. The half diallel where p parents are crossed in all possible combinations, parental selfs are included but that no reciprocals are included. These types of design will contain $p[p+1]/2$ families ($[p[p-1]/2]$ segregating populations of F_1s and p inbred parents).
- **Method 3**. The full diallel without parent selfs, which consists of all cross combinations (including reciprocals) of p parents. Method 3 differs from Method 1 in that with Method 3 the inbred parents are not included in the diallel design.
- **Method 4**. The half diallel, without parent selfs, which consists of all p parents crossed in all possible combinations (but with no reciprocals). The

Method 4 design differs from the Method 2 design as the inbred parents are not including in the Method 4 design.

Griffing's Analysis allows the option to test for **fixed** (*Model 1*) or **random** (*Model 2*) effects. Fixed effect models are where inference is made only on the parents that are included in the diallel cross while random effects models are where inference is made regarding all possible parents from a crop species. Therefore, in fixed effect models the parents used in the diallel cross are specifically chosen (i.e. because a breeder wishes to have additional information regarding general or specific combining ability of chosen lines). In random effects models the parental lines should be chosen completely at random. If this is done then the analyses can be interpreted to cover the eventuality that *any parents are used*.

Obviously, in most cases where plant breeders are involved, it is often very difficult to decide whether the parental lines were *chosen* or identified at *random*. In many cases the parents in diallel crossing designs are a sample of already commercial cultivars. In this case some would argue that being commercial cultivars they cannot be a random sample, as by definition all commercial cultivars are a very narrow subset of all potential genotypes within a species. On the other hand, others have argued that plant breeders are only interested in genotypes of commercial or near-commercial standard and they can therefore quite rightly term their choice as a random sample of commercially suitable cultivars.

There are no hard and fast rules regarding fixed or random models, and usually there is little to be lost or gained from either argument, provided that the analyses are not treated as one type and interpreted as another. For example, plant breeders and researchers often include diverse parental genotypes as parents in diallel crossing designs (and we believe this to be an excellent idea). However, do not choose specific parental lines which show a range of expressions for (say) yielding ability, cross them in a diallel design, and try to infer from the results what would happen if any different lines were included.

Griffing's Analysis requires no genetic assumptions and has been shown by many researchers to provide reliable information on the combining potential of parents. Once identified the 'best' parental lines (those with the highest general combining ability) can be crossed to identify optimum hybrid combinations or to produce segregating progeny from which superior cultivars would have high frequency.

In simplest terms, the cross between two parents (i.e. parent $i \times$ parent j) in Griffing's Analysis would be expressed as:

$$X_{ij} = \mu + g_i + g_j + s_{ij}$$

where μ is the overall mean of all entries in the diallel design, g_i is the general combining ability of the ith parent, g_j is the general combining ability of the jth parent and s_{ij} is the specific combining ability between the ith parent and the jth parent.

General combining ability (GCA) measures the average performance of parental lines in cross combination. GCA is therefore related to the proportion of variation that is genetically additive in nature.

Specific combining ability (SCA) is the remaining part of the observed phenotype that is not explained by the general combining ability of both parents that constituted the progeny.

Griffing's Analysis of a diallel is by analysis of variance, where the total variance of all entries is partitioned into: general combining ability; specific combining ability and error variances. In cases where reciprocals are included, then reciprocals (or maternal effects) are also partitioned. Error variances are estimated by replication of families. To avoid excessive repetition, only Method 1 (complete diallel) and Method 2 (half diallel) both including parents will be considered further.

Degrees of freedom (df), sum of squares (SS) and mean squares (MSq) from the analysis of variance for Method 1 for the assumption of model 1 (fixed effects) are shown in Table 6.1. Also shown are the expectations for the mean squares (EMS).

Similar expected mean squares for Method 1, model 2 (random effects) are shown in Table 6.2.

Considering now Method 2 (the half diallel), the degrees of freedom (df), sum of squares (SS), mean squares (MSq) and expected mean squares (EMS) for

Table 6.1 Degrees of freedom, sum of squares and mean squares from the analysis of variance of a full diallel including parent selfs (Method 1) assuming fixed effects. Also shown are the expectations for the mean squares.

Source	df	SS	MSq	EMS
GCA	$p-1$	S_g	M_g	$\sigma^2 + 2p(1/(1-p))\Sigma g_i^2$
SCA	$p(p-1)/2$	S_s	M_s	$\sigma^2 + 2/(p(p-1))\Sigma_{ij}s_{ij}^2$
Reciprocal	$p(p-1)/2$	S_r	M_r	$\sigma^2 + 2(2/(p(p-1)))\Sigma_{i<j}r_{ij}^2$
Error	$(r-1)p^2$	S_e	M_e	σ^2

Table 6.2 Degrees of freedom, sum of squares and mean squares from the analysis of variance of a full diallel including parent selfs (Method 1) assuming random effects. Also shown are the expectations for the mean squares.

Source	df	SS	MSq	EMS
GCA	$p-1$	S_g	M_g	$\sigma^2 + 2p(1/(1-p))\sigma_s^2 + 2p\sigma_g^2$
SCA	$p(p-1)/2$	S_s	M_s	$\sigma^2 + 2((p^2-p+1))/p^2\sigma_s^2$
Reciprocal	$p(p-1)/2$	S_r	M_r	$\sigma^2 + 2\sigma_r^2$
Error	$(r-1)p^2$	S_e	M_e	σ^2

For Method 1, where r is the number of replicates; p is the number of parents; S_g is $1/2p\Sigma_i(X_{i.} + X_{.i})^2 - 2/p^2 X_{..}^2$; S_s is $1/2\Sigma_{ij}x_{ij}(x_{ij} + x_{ji}) - 1/2p\Sigma_i(X_{i.} + X_{.i})^2 + 1/p^2 X_{..}^2$; S_r is $1/2\Sigma_{i<j}(x_{ij} - x_{ji})^2$ and $X_{i.}$ is $\Sigma_j x_{ij} = x_{i1} + x_{i2} + x_{i3} + \cdots$, that is, sum over rows; $X_{.j}$ is $\Sigma_i x_{ij} = x_{1j} + x_{2j} + x_{3j} + \cdots$, that is, sum over columns and $X_{..}$ is $\Sigma_{ij}x_{ij}$ is sum of all observations.

Table 6.3 Degrees of freedom, sum of squares and mean squares from the analysis of variance of a half diallel including parent selfs (Method 2) assuming fixed effects. Also shown are the expectations for the mean squares.

Source	df	SS	MSq	EMS
GCA	$p-1$	S_g	M_g	$\sigma^2 + (p+2)(1/(1-p))\Sigma g_i^2$
SCA	$p(p-1)/2$	S_s	M_s	$\sigma^2 + 2(p/(p-1))\Sigma_j s_{ij}^2$
Error	$(r-1)\{p(p+1)/2\}$	S_e	M_e	σ^2

Table 6.4 Degrees of freedom, sum of squares and mean squares from the analysis of variance of a half diallel including parent selfs (Method 2) assuming random effects. Also shown are the expectations for the mean squares.

Source	df	SS	MSq	EMS
GCA	$p-1$	S_g	M_g	$\sigma^2 + \sigma_s^2 + (p+2)\sigma_g^2$
SCA	$p(p-1)/2$	S_s	M_s	$\sigma^2 + \sigma_s^2$
Error	$(r-1)[p(p+1)/2]$	S_e	M_e	σ^2

Where r is number of replicates; p is number of parents; S_g is $1/(p+2)\{\Sigma_i(X_{i.}+x_{ii})^2 - 4/pX_{..}^2\}$; S_s is $\Sigma_{i<j}x_{ij}^2 - 1/(p+2)\Sigma_i(X_{i.}+x_{ii})^2 + 2/((p+1)(P+2)X_{..}^2)$ and $X_{i.}$ is $\Sigma_j x_{ij} = x_{i1}+x_{i2}+x_{i3}+\cdots$, that is, sum over rows; $X_{..}$ is $\Sigma_{ij}x_{ij} = $ is sum of all observations.

model 1 are shown in Table 6.3 and Method 2 and model 2 in Table 6.4.

When SCA is relatively small in comparison to GCA it should be possible to predict the performance of specific cross combinations based only on the values obtained for GCA of parents. A relatively large SCA/GCA variance implies the presence of dominance and/or epistatic gene effects. It should also be noted that if dominance × additive effects are present, the GCA component will also contain some of these effects in addition to pure additive effects.

For inbred lines, the closer that the following equations are equal to one (i.e. as SCA becomes small or very small compared to GCA), then greater predictability based on GCA will be possible. The ratio equations for each model are:

$$\text{Model 1: } 2g_i^2/[2g_i^2 + s_{ij}^2]$$

$$\text{Model 2: } 2\sigma_g^2/[2\sigma_g^2 + \sigma_s^2]$$

where g_i^2, σ_g^2 are the general combining ability mean square and variance, respectively and s_{ij} and σ_s^2 are specific combining ability mean square and variance, respectively.

The choice of Griffing method will depend on the plant breeder or researcher's preference and on the characters of the crop and trial under investigation. If, for example, there is a suspicion that the particular inheritance has a maternal or cytoplasmic effect then Method 1 or Method 3 may be the desired choice. If, however, there is no evidence of reciprocal differences then Method 2 or Method 4 would be chosen. When the variance components are of major importance then it has been suggested that Method 1 will result in a more accurate and constant variance estimation compared to the other methods available. Conversely, it has been reported that the inclusion of the parental genotypes in the diallel design can cause an upward bias in the estimation of the GCA and SCA variances.

Normally the F_1 generation is considered in Griffing's Analysis. However, as no genetic assumptions are involved then there are no reasons why F_2 or indeed other segregating generations could not be analyzed.

Despite the attraction and simplicity of Griffing's Analysis several researchers have criticized the technique.

Table 6.5   Average plant height of each of the 45 F_1's and the 10 parents in a half diallel with selfs.

	Global	Helios	Jaguar	Starr	93.C.3	Westar	DNK.89.213	Cyclone	Hero	Reston
Global	328									
Helios	341	352								
Jaguar	336	310	263							
Starr	329	271	293	287						
93.C.3	269	308	271	312	292					
Westar	256	350	279	299	324	273				
DNK.89.213	284	313	290	266	259	270	293			
Cyclone	321	263	280	241	285	243	273	201		
Hero	246	261	315	261	241	256	250	244	231	
Reston	306	327	295	287	284	296	275	265	248	277
GCA	+18.4	+26.4	+9.4	+0.4	+0.4	+0.4	−6.6	−22.6	−28.4	+2.4

In open-pollinated species, where GCA is the only parameter of interest, then it has been suggested that other designs such as topcross or polycross would yield equally reliable results with less effort and that these alternative methods provide the opportunity to test many more parental lines. Similarly it has been argued that in many instances North Carolina I designs (where a set of p parents to be tested are each inter-crossed with a set number of other parents and where each parent under test is not necessarily crossed to the same tester) or North Carolina II designs (where a set of p parents are crossed to a common set of n different parents and where each parent under test is crossed to the same set of non-test parental (or tester) lines) would offer a better alternative to diallel designs and Griffing's Analysis.

Many studies have shown that the GCA values of parents from diallel analyses are similar to actual phenotypic performance of the parents. It has, therefore, been argued that it is not necessary to progeny test potential parents in a plant breeding programme but simply to '*cross the best with the best*'. Many practical plant breeders often add to this statement, however, '*cross the best with the best, and hope for the best*', but perhaps that is what we would be doing anyhow.

Example of Griffing analysis of half diallel

Let us consider now an example of a half diallel. A half diallel crossing design between ten homozygous lines of spring canola (*Brassica napus*) was carried out in the spring of 1992. The parental lines were: Global, Helios, Jaguar, Starr, 93.C.3.1, Westar, DNK.89.213, Cyclone, Hero and Reston. Hero and Reston are both industrial

rapeseed cultivars while the others are canola (edible) types. Crossing resulted in $n[n − 1]/2 = 45$ different F_1 families. Over the following winter each of the 45 F_1 families were grown in a two replicate randomized complete block design which also included the 10 parent selfs making a design with 55 entries ($n[n + 1]/2$) and two replicates (i.e. 110 plots).

Throughout the growth of this experiment a number of different traits were recorded on each of the 110 plots. To avoid excessive repetition we will only consider one of these characters, plant height at end of flowering.

The average plant height of each of the 45 F_1s and the 10 parents are shown in Table 6.5. The data used were the sum of two plant heights (cms) as two readings were made on each of the replicate plots.

From the data the total variance (sum of squares) is partitioned into differences between the two replicate blocks (Reps), general combining ability, specific combining ability and an error term (based on interactions between replicates and other factors). Sum of squares (SS) and mean squares (MS) obtained are shown in Table 6.6.

The basic assumption of this experiment was that the ten parental lines were ***chosen*** as representative of the wide range of *B. napus* cultivar types that were available. We are therefore analyzing a ***fixed effect model*** and all the mean squares in the analysis are tested for significance (using the 'F' test) against the error mean square (i.e. 1545).

From the analysis the overall replicate block effect (i.e. difference between replicate one and replicate two) was not significant. An F-value is obtained for specific

Table 6.6 Degrees of freedom, sum of squares and mean squares from the analysis of variance of plant height of a half diallel including parent selfs. In the analysis the total variance is partitioned into differences between the two replicate blocks (Reps), general combining ability, specific combining ability and an error term (based on interactions between replicates and other factors).

Source	df	Sum of squares	Mean squares
General combining ability	9	108 665	12 074
Specific combining ability	45	113 497	2522
Replicate blocks	1	959	959
Replicate Error	54	83 428	1545
Total	109	306 548	2812

combining ability by $2522/1545 = 1.63$. This 'F' value is compared to F-values found in statistical tables at differing probability levels and with 45 and 54 degrees of freedom. When this is done, with some degree of difficulty, it is found that the probability of this F value occurring if SCA were not significant is 95.7, therefore specific combining ability is just significant at the 5% level.

Consider now the variance ratio for general combining ability. The appropriate F-value is $12\,074/1545 = 7.8$. When this value is compared to the appropriate F-values in statistical tables with 9 and 54 degrees of freedom we find that it exceeds the appropriate expectation based on 99.9% confidence (i.e. approximately 3.54) and so we say that general combining ability is highly significant. This, in combination with the marginal significant of specific combining ability, suggests an additive–dominance model with high additive effects.

Now the expected mean square for specific combining ability of a half diallel and fixed effects is:

$$\sigma^2 + 2(p/(p-1))\Sigma_i s_i^2$$

Therefore

$$2521 - 1545 = 2(10/(10-1))\Sigma_i s_i^2$$
$$976 = 2.2\Sigma_i s_i^2$$

so

$$\Sigma_i s_i^2 = 976/2.2 = 439$$

Similarly for general combining ability, the expected mean square is:

$$\sigma^2 + (p+2)(1/(1-p))\Sigma g_i^2$$

Therefore

$$12\,074 - 1545 = (10+2)(1/(1-10))\Sigma g_i^2$$
$$10\,529 = 1.33\Sigma g_i^2$$

so

$$\Sigma g_i^2 = 10\,529/1.33 = 7897$$

Now, from the equation above we can compare GCA and SCA effects, as noted earlier we have:

$$2g_i^2/[2g_i^2 + s_{ij}] = 2 \times 7896.893/$$
$$[(2 \times 7896.893) + 439.288]$$
$$= 0.973$$

As this value is very close to one, it indicates that s_{ij}^2 is relatively small compared to g_i^2. Therefore additive genetic effects predominate. This means there is a good chance that plant height at the F_1 stage in a *B. napus* breeding programme can be predicted with good accuracy depending on the general combining ability of chosen parental lines.

In many instances there is good agreement between the general combining ability of a genotype and the phenotypic performance of the line. If this is the case then all that is necessary is to determine the expression of the parents and from these the expected expression of the offspring can be estimated (compare with h_n^2). Therefore in this example consider the regression of average parental performance against the offspring (Figure 6.1). It can be clearly seen that there is relatively good agreement between parents and offspring. The regression equation is offspring mean $=0.7265 \times$ parent $+80.0$. Therefore the narrow sense heritability of these data is approximately 0.73, which is relatively high in that 73% of the total variation is additive genetic variance.

Hayman and Jinks' analysis

Hayman and Jinks developed an analysis for diallels that has been widely used by many plant researchers to evaluate the mode of inheritance. This analysis is based on a model that, for any one locus, i, with two alleles, the difference between the two homozygotes is $2a$.

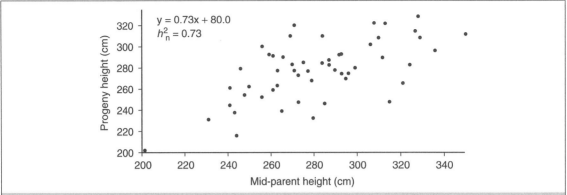

Figure 6.1 Scatter diagram of mid-parent phenotype height against average offspring progeny phenotype height from a 10 × 10 half diallel in *Brassica napus*.

The difference between the heterozygous (F_1) and the mid-parent value (m) is d.

To simply interpret a Hayman and Jinks' Analysis the following assumptions are made:

- Diploid segregation
- Homozygous parents
- No difference between reciprocal crosses
- No epistasis
- No multiple alleles
- Genes are distributed independently between the two parents

But these assumptions are tested in the approach.

The parents and all possible F_1 progenies are evaluated for the trait of interest. All the offspring of one parent used in crosses is called an **array**. That is, in all crosses that the particular parent was used. Seven kinds of variances and covariances are calculated including:

V_p = variance among the parent lines;

V_r = the variance among family(F_1 and reciprocal) means within an array;

V_{xr} = variance among the means of the arrays;

$\bar{V}_r$ = mean value of all V_r over all arrays;

W_r = the covariance between families within the ith array and their non-recurrent parent;

$\bar{W}_r$ = mean value of W_r over all arrays;

σ_E^2 = Error variance.

From these, a number of parameters can be estimated, including:

$$V_A = 4/7[V_p + \bar{W}_r + V_{xr}] - \sigma_E^2$$

$$V_D = 4\bar{V}_r - V_A$$

The estimates of V_A and V_D indicate the amounts of additive variance and dominance variance among the crosses. This estimate of V_D assumes that F_1 progeny are being evaluated (although other generations can be accommodated). Obviously the frequency of heterozygous alleles in a population will determine the degree of dominance variation and this will vary with successive rounds of selfing.

The most useful aspect of Hayman and Jinks' Analysis for plant breeders involves examination of variance and covariance relationships and estimation of V_A or V_D. Therefore we will only cover the within array variances and between array covariances, how they can help in determining the inheritance of the character of interest, what the relationship of these two parameters means in comparing different parental lines, and estimation of h_n^2.

Based on the assumptions (listed above) of Hayman and Jinks analyses we have

$$\bar{V}_r = \frac{1}{4}(V_A + V_D)$$

$$\bar{W}_r = \frac{1}{2}V_A$$

Consider now the relationship between W_r and V_r. If we plot W_r against V_r, the regression line must have a slope which will pass through the point $(\bar{V}_r, \bar{W}_r)$ and will have expected value of one only if the additive–dominance model is adequate to explain the variation observed. It should, therefore be noted that the relationship between W_r and V_r provides a test of the additive–dominance model of gene action. If the contribution of many genes are not independent, that is if there are genes interacting in their effect (epistasis) we would not expect the relationship between W_r and V_r to be as described. Hence if the additive–dominance model is not adequate then the regression of W_r against V_r will not result in a regression slope of one.

In addition, regression of W_r against V_r will result in a gradient which will pass through the point $\frac{1}{4}(V_A + V_D), \frac{1}{2}V_A$ and which will cut the y-axis at $\bar{W}_r - \bar{V}_r = \frac{1}{2}V_A - \frac{1}{4}(V_A + V_D) = \frac{1}{4}(V_A - V_D)$. So we can learn something about the average dominance relationships of the quantitative inheritance system. If additive genetic variance (V_A) is greater than dominance genetic variance (V_D) then the regression line will cut the y-axis (the W_r-axis) above zero. Similarly, the reverse will be true of V_D is greater than V_A.

The relative position of each array $(V_r$ and $W_r)$ will indicate the relative frequency of the dominant to recessive alleles that array parent has. Therefore, the relative position of the array points on the line will reflect the direction of dominance. If an array has a scatter of points close to the origin (i.e. a low V_r and W_r values) this indicates that the parent common to that array has a high frequency of dominant alleles for that character of interest. If an array has a scatter of points at a distance from the origin (i.e. high V_r and W_r values) then the common parent in that progeny array will have a relatively high frequency of recessive alleles.

This graph (W_r/V_r) can therefore provide a great deal of information about the genetic situation between the parents in the diallel. In plant breeding terms the frequency of dominant (or recessive) alleles, combined with the average progeny performance can be useful indicators for selection. For example, given two possible parents, if one has a high frequency of recessive alleles and the other a high frequency of dominant alleles for, say, yield. If both parents have similar general combining ability, then a plant breeder should choose the recessive parent as it will be easiest to select and fix for high yield. Similarly, selection for high yield, which is related to a high frequency of dominant alleles will

likely have lower narrow-sense heritability compared to the case of high recessive allele frequency.

To illustrate further, consider the example from a half diallel design involving 10 homozygous parents of spring canola/rapeseed (*B. napus*). Although many traits have been recorded from this trial we will again consider plant height which was explained as an example of the Griffing's Analysis earlier. The means, over replicates, have therefore been shown earlier. From the array means, values of within array variances (V_r) and covariances (W_r) were obtained. Similar V_r and W_r values were calculated from each of the two replicates.

Regression analysis of V_r against W_r resulted in a regression equation:

$$W_r = 0.60 \times V_r + 59.07$$

The analysis of regression resulted in a mean square for linearity of 47 473 and a mean square for departure from linearity to be 9 286. From this we calculate an 'F' value of 5.11 which is significantly $(p < 0.05)$ larger than would have been expected if the relationship between W_r and V_r was not linear.

The standard error (se_b) of the regression coefficient (b) was 0.265. From this we can calculate the Student's t as:

$$t = \frac{b-1}{se_b} = \frac{0.400}{0.265} = 1.51$$

This did not exceed the value from t-tables with 8 degrees of freedom $(n-2)$ and $p < 0.05$. Therefore **b** is not significantly different from a regression slope of one and so we have a good indication that the additive–dominance mode is adequate to explain the inheritance of plant height in *B. napus*.

The regression line cuts the y-axis at $(V_r = 0)$ above the origin $(+59.07)$ so we can say that additive effects are greater than dominance effects.

A scatter diagram of the W_r and V_r values from the 10 arrays (parents) is shown in Figure 6.2. From the diagram we see that the cultivars Hero, DNK.89.213, Jaguar and 93.C.3 are relatively close to the origin while Helios, Cyclone and Westar are further from the origin. From this we can deduce that those closest to the origin have a higher frequency of dominant alleles for plant height and those further from the origin have highest frequency of recessive alleles for plant height. In the extremes, the cultivar Hero has highest relative frequency of dominant alleles and the cultivar Helios has highest frequency of recessive alleles for plant height.

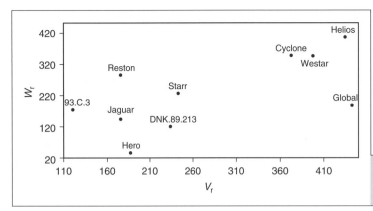

Figure 6.2   Scatter diagram of V_r and W_r values from arrays in a 10×10 half diallel in *Brassica napus*.

Moving on to the $W_r + V_r$ and $W_r - V_r$ from each array. $W_r + V_r$ and $W_r - V_r$ will contain all the information that W_r and V_r contain. Now if dominance is present then $W_r + V_r$ will vary from array to array. If there is non-allelic interaction then $W_r - V_r$ will vary from array to array. If only dominance is present then $W_r - V_r$ will not vary between arrays more than would be expected by sampling variation. We can calculate the values of $W_r + V_r$ and $W_r - V_r$ from the V_r and W_r values obtained from each replicate and carry out a one-way analysis of variance on the resulting data. When this is done we have the two analyses of variance tables:

$$W_r + V_r$$

Source	df	MSq	F-value	Significance
Between arrays	9	17 164	4.63	$(0.01 < p < 0.05)$
Within arrays	10	3708		

$$W_r - V_r$$

Source	df	MSq	F-value	Significance
Between arrays	9	5642	2.60	n.s.
Within arrays	10	2170		

Therefore values of $W_r + V_r$ vary significantly between arrays ($p < 0.05$) so we can say that dominance is present. Values of $W_r - V_r$ between different arrays are not significantly different and therefore we can say that there is no evidence of non-allelic interaction and that only dominance is present.

These data are from homozygous parents and F_1 progenies. In many cases it is difficult to obtain large quantities of F_1 seed and the actual diallel analysis needs to be carried out on the F_2 (or higher) generations. When this is done the same six assumptions listed at the beginning of this section still apply. The regression of W_r / V_r will still have an expected slope of unity if the additive-dominance model is adequate to explain the inheritance, and $W_r - V_r$ should be constant across arrays if no epistasis is present. Therefore, as far as the example above is concerned, then it would make no difference if F_2 data were used. However, if a more detailed analysis is to be carried out and the components V_A and V_D are to be estimated then some modification is needed. The modification is not within the scope of this book.

Estimating h_n^2 from Hayman and Jink's Analysis

If the crop under investigation in a diallel crossing design complies with all the restraints of the Hayman and Jinks design, then it is possible to obtain accurate estimates of additive genetic variance (V_A) and dominant genetic variance (V_D) straightaway and hence determine the narrow-sense heritability (h_n^2). The average W_r value ($\overline{W}_r$) is an estimate of $\frac{1}{2}V_A$. The V_p value is a direct estimate of V_A, and the V_{xr} value is an estimate of $\frac{1}{4}V_A$. These relationships hold true irrespective of the generation (i.e. F_1, F_2, F_3, etc.) that is analyzed. From these three estimates of V_A, we can produce a weighted mean where:

$$V_A = 4/7[V_p + \bar{W}_r + V_{xr}]$$

The dominance genetic variance (V_D) will vary from generation to generation. Greatest V_D will be observed in the F_1 generation as there is greatest frequency of heterozygotes compared to other generations. In F_1 the average V_r value ($\bar{V}_r$) is an estimate of $\frac{1}{4}[V_A + V_D]$, and V_D can easily be estimated by substituting the already calculated V_A value in to this equation. Therefore, when analyzing data from F_1 family diallels:

$$D = 4\bar{V}_r - V_A$$

From estimation of V_A and V_D we can now calculate h_n^2:

$$h_n^2 = \frac{1}{2}V_A / \left[\frac{1}{2}V_A + \frac{1}{4}V_D + \sigma_E^2\right]$$

where, σ_E^2 is the replicate error term obtained from the analysis of variance in the *B. napus* example shown earlier in the Griffing's Analysis.

In F_2 families $V_D = \frac{1}{4}[V_A + \frac{1}{4}V_D]$ and so $V_{D-F_2} = 16\bar{V}_r - 4V_A$, and in F_3 families, $V_{D-F_3} = \frac{1}{4}[A + 1/16 V_D]$. It should be noted that the proportion of V_D in each family is decreased each generation by $[\frac{1}{2}]^n$, where n is the generation number (i.e. $[\frac{1}{2}]^1 = 1$ at F_1; $[\frac{1}{2}]^2 = \frac{1}{4}$ at F_2; $[\frac{1}{2}]^3 = 1/16$ at F_3, etc.).

CROSS PREDICTION

There is one further way that it is possible to predict the response to selection, in the long-term, although not necessarily the rate of response. This approach is based on the genetics underlying the traits, was proposed by Jinks and Pooni, and is currently attracting considerable attention in terms of experimental investigations and in applying it to practical breeding. This will be covered in more detail in the next chapter but needs mentioning here to keep in view the options available to the breeder in terms of making predictions.

If, to start with, we assume that we have an inbreeding species and wish to produce a final variety that is true-breeding.

What we want to know of any population or cross is what is the distribution of inbred lines that we predict can be derived from it and findout what is the probability of one of these lines having a phenotype equal to, or exceeding, any target level that we set, in other words, that we would be aiming for with selection.

If we assume that the distribution of the final inbred lines that are derivable have a normal distribution as is generally the case in practice, then it can be described by the mean and standard deviation. Since they are inbred lines they will have a mean of m and a standard deviation of $\sqrt{V_A}$, we can predict the properties of the distribution of all inbred lines possible and hence we can obtain the frequency (= probability) of inbreds falling into a particular category. In other words, we can simply use the properties of the normal probability integral in tables to say what the probability of obtaining an inbred line with expression falling in a particular category. If the probability is low it will obviously be difficult to actually obtain such a line. If the probability is high it will be easy to produce.

How do we put it into practice? If we have a set of genotypes for use as Parents, which ones do we cross to produce our desired new inbred lines? Do we take A×B and C×D or A×D and C×Z etc.? We will need to decide between the crosses before we invest too much time and effort, otherwise we may well be spreading our efforts over crosses that will not produce the phenotypes we want. If we take the crosses and estimate m and V_A for each, then we can estimate the probability of obtaining our desired target values. From this we can rank the crosses on their probabilities and only then use the ones with the highest probabilities of producing lines with the required expression of characters deemed to be important.

In fact, the approach is even more general in that it can be used to predict the properties of the F_1 hybrids derived from the inbred lines. It can also be used to predict the probability of combination of characters, that is the probability of obtaining desirable levels of expression in a series of characters.

What are the drawbacks to the approach? First, we need to estimate m and A, and this involves a certain amount of work in itself, but is fairly modest.

Second, it also assumes that the estimates we use are appropriate to the final environment that the material is to be grown. In other words, as in the case of heritabilities, if we carry out the experiments in one environment at one site in one year, we are assuming that this is representative of other years and sites. We can, of course, carry out suitable experiments to obtain estimates in more years and sites, but this involves extra time and effort.

The use of cross prediction techniques in selection will be discussed in greater detail in Chapter 7.

THINK QUESTIONS

(1) Given values for the variance of the mean of the F_2, and variance of the F_1, estimate h_b^2 and explain what this tells us about the genetic determination of the trait.

$$V_{\bar{F}_2} = 436.72$$

$$V_{\bar{F}_1} = 111.72$$

Given below are the variances of the mean from two parents (P_1 and P_2), the F_1, F_2, and both backcross families (B_1 and B_2), estimate h_n^2 and explain what the value means in genetic terms.

$$V_{\bar{P}_1} = 14.1 \quad V_{\bar{P}_2} = 12.2$$

$$V_{\bar{F}_1} = 13.3 \quad V_{\bar{F}_2} = 40.2$$

$$V_{\bar{B}_1} = 35.2 \quad V_{\bar{B}_2} = 34.6$$

(2) List the six assumptions necessary for a straight forward interpretation of a Hayman and Jinks' Analysis of diallels.

Below are shown values of array means, within array variances (V_r) and covariances between array values and non-recurrent parents (W_r) from a Hayman and Jinks Analysis of a 7×7 complete diallel in dry pea.

Parent name	V_r	W_r	Array mean
'Souper'	34.1	19.3	456
'Dleiyon'	99.9	79.3	305
'Yielder'	21.0	11.2	502
'Shatter'	99.4	68.4	314
'Creamy'	49.6	39.4	372
'SweetP'	59.1	48.8	361
'Limer'	61.8	49.2	393

The variate of interest is pea yield. Regression of W_r against V_r resulted in the equation: $W_r = 0.837 \times V_r - 4.817$, with standard error of the regression slope equal to $se_b = 0.0878$. An analysis of variance of $W_r + V_r$ showed significant differences between arrays while a similar analysis of $W_r - V_r$ showed no significant differences between arrays. What can be deduced regarding the inheritance of pea yield from the information provided? If you were a plant breeder interested in developing high yielding dry pea cultivars, on which two parental lines would you concentrate your breeding efforts? Briefly explain why.

(3) Below is shown an analysis of variance of plant yield from a 6×6 half diallel (including parents). The analysis of variance is from a Griffing's Analysis (Model 2). GCA = general combining ability, SCA = specific combining ability, Error = random error obtained by replication, df = degrees of freedom and SS = sum of square.

Source	df	SS
GCA	5	4988
SCA	15	6789
Error	21	5412

Discuss the results from the analysis given that the 6 parents were: (1) specifically chosen and (2) chosen completely at random.

(4) It is desired to determine the narrow-sense heritability for flowering date in spring canola (*B. napus*). Both parents and their offspring from ten cross combinations were grown in a properly designed field experiment. At harvest, yield was recorded for each entry and using these data the average phenotype of two parents (i.e. $[P_1+P_2]/2$) was considered to be the x independent variable while the performance of their offspring was considered as the y dependant variable. A regression analysis is to be carried out by regression of the offspring (y) onto the average parent (x). The following data are derived: $SP(x, y) = 345.32$; $SS(x) = 491.41$; $SS(y) = 321.45$. Estimate the slope of the regression (**b**), test if this slope is

greater than zero and estimate the narrow-sense heritability from the regression equation.

How would the relationship between the regression and the narrow-sense heritability differ if the regression were carried out between only the male parent and the offspring?

(5) Four types of diallel can be analyzed using Griffing's Analysis. Describe these types. Families from a 5×5 half diallel (including selfs) were planted in a two replicate yield trial at a single location. The parents used in the diallel design were chosen to be the highest yielding lines grown in the Pacific-Northwest region. Data for yield were analyzed using a Griffing's Analysis of variance. Family means, averaged over two replicates, degrees of freedom and sum of squares (SS) from that analysis are shown below. Explain the results from the Griffing's Analysis. What differences would there be in your analytical methods if the parents used had been chosen at random.

	Parent 1				
Parent 1	62.0	Parent 2			
Parent 2	71.0	69.5	Parent 3		
Parent 3	55.5	52.5	50.5	Parent 4	
Parent 4	72.5	80.5	56.5	76.5	Parent 5
Parent 5	70.5	66.5	36.5	71.0	64.5

Source	d.f.	SS
GCA	4	6694.058
SCA	10	825.676
Replicates	1	8.533
Error	14	317.467
Total	29	7845.733

From the same diallel data (above), within array variances (V_r) and between array and non-recurrent parent covariances (W_r) were calculated. The values of V_r and W_r for each parent along with the of mean of V_r, mean of W_r, sum of squares if V_r (ΣV_r^2), sum of squares of W_r (ΣW_r^2) and sum of products $\Sigma V_r W_r$ are shown below.

	V_r	W_r
1	98.0	102.0
2	66.3	53.2
3	161.8	207.4
4	50.3	65.2
5	71.4	83.1

Mean of $V_r = 89.56$; Mean of $W_r = 102.19$; $\Sigma V_r^2 = 7702.01$: $\Sigma W_r^2 = 15192.05$; $\Sigma V_r W_r = 10535.29$. From these data, test whether the additive-dominance model is adequate to describe variation between the progenies. What can be determined about the importance of additive compared to dominance genetic variation in this study. From all the results (Griffing and Hayman and Jinks, above) which two parents would you use in your breeding programme and why?

(6) Two genetically different homozygous lines of canola (*B. napus* L.) were crossed to produce F_1 seed. Plants from the F_1 family were self-pollinated to produce F_2 seed. A properly designed experiment was carried out involving both parents (P_1 and P_2, 10 plants each), the F_1 (10 plants) and the F_2 families (64 plants) and was grown in the field. Plant height of individual plants (cm) recorded after flowering. The following are family means, variances and number of plants observed for each family.

Family	Mean	Variance	Number of plants
P_1	162	1.97	10
P_2	121	2.69	10
F_1	149	3.14	10
F_2	139	10.69	34

Complete a statistical test to determine whether an additive–dominance model of inheritance is appropriate to adequately explain the inheritance of plant height in canola. If the additive–dominance model is inadequate, list three factors that could cause the lack of fit of the model.

(7) Given values for the variance of the F_2, and variance of both parents (P_1 and P_2) and the F_1, estimate h_b^2 and explain what this value means in genetic terms.

$$\sigma_{F_2}^2 = 436.72; \quad \sigma_{F_1}^2 = 111.72$$

$$\sigma_{P_1}^2 = 164.13; \quad \sigma_{P_2}^2 = 109.33$$

Given the variance from two parents (P_1 and P_2), the F_1, F_2, B_1 and B_2 families, estimate the narrow-sense heritability (h_n^2) and explain what the value means in genetic terms.

$$\sigma_{P_1}^2 = 9.5; \quad \sigma_{P_2}^2 = 7.4$$

$$\sigma_{F_1}^2 = 8.6; \quad \sigma_{F_2}^2 = 17.7$$

$$\sigma_{B_1}^2 = 14.3; \quad \sigma_{B_2}^2 = 15.2$$

(8) A new oil crop (*Brassica gasolinous*) has been discovered which may have potential as a renewable biological fuel oil substitute. This diploid species is tolerant to inbreeding and is self-compatible. A preliminary genetic experiment was designed to examine the inheritance of seed yield (YIELD) and percentage oil content (%OIL). This experiment involved a 4 × 4 half diallel (including selfs). The four homozygous parental lines are represented by the codes AAA, BBB, CCC and DDD. The half diallel array values (averaged over two replicates), array means, general combing ability (GCA) values, mean squares from the analyses of variance (Griffing style), V_r and W_r values (Hayman and Jinks' analysis), variance of array means (V_{xr}) and parental variances (V_p), and the one-way analyses of variance for $V_r + W_r$ and $V_r - W_r$ are shown below for each character.

Yield					%Oil			
AAA	40.5				20.5			
BBB	38.5	29.5			20.5	25.0		
CCC	37.0	28.0	19.5		23.0	26.0	30.5	
DDD	32.5	20.5	18.5	10.0	24.5	27.5	31.0	36.0
	AAA	BBB	CCC	DDD	AAA	BBB	CCC	DDD
Array means	37.1	29.1	25.8	20.4	22.1	24.7	27.6	29.8

Source	df	Yield	%Oil
G.C.A.	3	796.5	180.6
S.C.A.	6	90.5	30.1
Replicate blocks	1	0.4	0.1
Replicate error	9	51.5	5.7

V_r and W_r values

	Yield		%Oil	
	V_r	W_r	V_r	W_r
AAA	46.2	171.3	15.6	50.7
BBB	218.2	376.7	36.3	75.0
CCC	297.7	436.7	58.3	98.0
DDD	344.3	472.3	97.3	132.3

V_{xr} and V_p values

	Yield	%Oil
V_{xr}	196.9	44.4
V_p	687.6	180.7

Source	df	Yield		%Oil	
		$V_r + W_r$	$V_r - W_r$	$V_r + W_r$	$V_r - W_r$
Between	3	8760	28.0	628	7.68
Within	4	133	17.0	115	4.77

Without using regression, estimate the narrow-sense heritability (h_n^2) for seed yield, and explain this value in genetic variance terms. Explain the analyses for Yield and outline any conclusions that can be drawn for these data. Explain the analysis for %Oil (percentage of seed weight that is oil) and outline any conclusions that can be drawn from these data. Which **one** of these four genotypes would you choose as a parent in your breeding programme? Explain your choice. Describe any difficulties suggested from these analyses in a breeding programme designed for selecting lines with high yield and high percentage of oil.

(9) F_1, F_2, B_1 and B_2 families were evaluated for plant yield (kg/plot) from a cross between two homozygous spring wheat parents. The following variances from each family were found:

$$\sigma_{F_1}^2 = 123.7; \quad \sigma_{F_2}^2 = 496.2$$

$$\sigma_{B_1}^2 = 357.2; \quad \sigma_{B_2}^2 = 324.7$$

Calculate the broad-sense (h_b^2) and narrow-sense (h_n^2) heritability for plant yield. Given the heritability estimates you have obtained, would you recommend selection for yield at the F_3 in a wheat breeding programme, and why?

(10) Griffing has described four types of diallel crossing designs. Briefly outline the features of each Method 1, 2, 3 and 4. Why would you choose Method 3 over Method 1? Why would you choose Method 2 over Method 1?

A full diallel, including selfs was carried involving five chickpea parents (**assumed to be chosen as fixed parents**), and all families resulting were evaluated at the F_1 stage for seed yield. The following analysis of variance for general combining ability (GCA), specific combining ability (SCA) and reciprocal effects (Griffing's Analysis) was obtained:

Source	d.f.	MS
GCA	5	30 769
SCA	10	10 934
Reciprocal	10	9638
Error	49	5136

Complete the analysis of variance and explain your conclusions from the analysis. Given that the parents were chosen at random, how would this change the results and your conclusions?

Plant height was also recorded on the same diallel families and an additive-dominance model found to be adequate to explain the genetic variation in plant height. Array variances V_rs and non-recurrent parent covariances (W_rs) were calculated and are shown alongside the general combining ability (GCA) of each of the five parents, below:

Parent	V_r	W_r	GCA
1	491.4	436.8	−0.76
2	610.3	664.2	+12.92
3	302.4	234.8	−14.32
4	310.2	226.9	−15.77
5	832.7	769.4	+17.93

Without further calculations, what can be deduced about the inheritance of plant height in chickpea?

(11) A 4×4 halfdiallel design (with selfs) was carried out in cherry and the following fruit yields of each possible F_1 family were observed:

	Small reds			
Small reds	12	Big yields		
Big yields	27	36	Jim's delight	
Jim's delight	21	35	27	Jacks' best
Jack's best	28	27	26	21

From the above data, determine the narrow-sense heritability for yield in cherry.

7

Selection

INTRODUCTION

Selection of all living organisms has been going on since life was first created. Natural selection (i.e. evolution) has resulted in the diversity of plant and animal life which exists today. All selection results in a change of gene frequencies. Throughout evolution, species have been changing, '*more fit*' genotypes have predominated while those which are less fit in regard to survival, have become extinct. The aim of plant breeding is to direct selection towards increasing the frequency of desirable gene combination which best suit agricultural systems.

In order to be successful in a selection programme two criteria need to be satisfied, being:

- There is variation between plants within the unselected population and the breeders must be able to distinguish between different phenotypes
- At least some of that variation must be genetic in nature

Obviously, if a plant breeder cannot distinguish any differences between plants within a population (or different populations) then it will be impossible to select those individuals which appear superior. Second, if the variation observed between plants within a population is the result purely of the environmental response of lines, with no genetic component, then there will be no progress made in a selection scheme.

WHAT TO SELECT AND WHEN TO SELECT

Having decided that the two criteria above are indeed satisfied, among the first tasks to be addressed by a plant breeder are to decide *what characters are to be selected for and at what stage in the breeding scheme will selection be applied.*

Consider the first question of *what to select for?* To address this a plant breeder must refer to the **breeding objectives**. These will have been set according to criteria such as:

- The potential market size of the crop
- The region targeted for propagation
- The major deficiencies which exist within cultivars which are presently available
- The economic implications of addressing deficiencies such as disease and pest resistance
- Needs of the farmer, such as rapid establishment, early maturity, plant height, resistance to lodging, harvest ability
- The need of the end-user, including: appearance; storability; processing quality; etc.

Many more factors may need to be included in setting the breeding objectives and the above list only mentions but a few of the more important questions which need to be addressed.

It is not usually possible to select for the wide range of characters needed for a successful new cultivar in a single season. Plant breeders therefore screen plant populations over several years, sometimes addressing a number of different traits at each evaluation stage. Having decided what characters are to have greatest priority, it is necessary to follow an organized scheme of selection to determine which characters will be addressed at the various stages.

The inheritance of traits will be of great importance in determining, not only whether selection is to be carried out, but also the complexity of experimentation needed in order to identify the desirable types.

Qualitative trait selection

Characters whose expression shows a qualitative form of inheritance can be easily selected for, provided that a suitable screening method is available to determine the presence or absence of the single gene in plants in seed-propagated crops. If the expression of the qualitative character is determined by a recessive allele, then a single round of selection should ensure that all selected plants are fixed for the particular trait. If the desirable allele is completely dominant, several rounds of recurrent selection will be necessary to ensure that the character is genetically fixed in selected plants.

Qualitative characters can often be selected relatively quickly and using very small plots (sometimes, even single plant plots) compared to quantitative inherited traits. The ease of selecting for single gene traits has resulted in these characters having high selection intensity in the early generation stages where most genotypes are evaluated and where it may only be possible to grow small plots.

Selection for such qualitative expression can indeed be a powerful tool in reducing the number of genotypes selected in a plant breeding scheme, although it should never be forgotten that it is often the quantitatively inherited characters which add greatest value to a new cultivar (i.e. yield, quality and durable plant resistance). If early generation selection is to be carried out for single gene traits, then the breeder must be sure that this selection is not having an adverse effect on the selected populations (i.e. no linkage between advantageous qualitative traits and adverse quantitative traits or any unwanted non-allelic interactions, or pleitropic effects).

Quantitative trait selection

Quantitatively inherited characters usually are more difficult to evaluate due to the higher potential for modification of expression by the environment. Greater experimentation (replication or plot size) is necessary to maximize selection response. As a result, many of the quantitative traits are not positively selected for in the early generation selection stages. Selection for these characters is often delayed until the numbers of genotypes which require testing are reduced and where greater amounts of planting material are available for

more sophisticated tests. For example, it is common practice in most plant breeding schemes not to select the early generation lines for quality traits which involve either large quantities of produce, which provide only crude estimates of worth with small samples or that are expensive.

Obviously any character which is considered of high importance should be selected for at the earliest stages of a plant breeding scheme where greatest variation will exist among families or populations but where the trial designs and amount of material make selection effective. Despite the simplicity of this statement, in practice it is often completely ignored.

The characters which are evaluated at different stages of a plant breeding scheme will be discussed in later sections.

Positive and negative selection

Two forms of selection are said to be available to plant breeders, positive and negative selection. It is difficult to clearly define the difference between the two types (and indeed, some wise and worldly breeders do not distinguish between the two). In simple terms negative selection is where the very worst plants or families are discarded while positive selection is where the very best plants or families are selected. Perhaps the simplest description would be related to the proportion of plants that are selected from a population. If more than 50% of the original population is selected then this can be considered negative selection. If less than 50% of the population is retained then this would be positive selection.

RESPONSE TO SELECTION

It has already been stated that selection will only be successful if there is sufficient phenotypic variation and that at least some of this variation is genetic in origin. It should be of no surprise, therefore, that the response to selection is related to heritability. Indeed consider the equation:

$$X_1 - X_{n-1} = R = i\sigma h^2$$

where X_1 is the mean phenotype of the selected genotypes, X_{n-1} is the mean phenotype of the whole population, R is the advance as a result of one round of selection, h^2 is the appropriate heritability (narrow-sense heritability for inbreeding crops or broad-sense heritability for out breeding crops), σ is the phenotypic standard deviation of the whole population and i is the *intensity of selection*, which is a statistical factor dependent upon the proportion of the population selected. The above equation is probably the most fundamental equation in plant breeding and should be kept in mind.

The intensity of selection (i) is related to the percentage of the population that is selected (k), and takes the values:

Percentage selected (k)	i
1	2.665
5	2.063
10	1.755
20	1.400

Although the intensity of selection (i) has been extensively tabulated for a range of different selection rates, in cases where the initial population is large (i.e. greater than 50 genotypes) and the proportion of genotypes selected less 20%, then the following equation can provide an estimate of i:

$$i = 0.77 + 0.96 \times \log(1/k)$$

From the tabulated values of selection intensities and the estimation equation it can be seen that there is not a linear relationship between higher selection rates (k) and greater response from selection. Retaining 10% of the selected population results in an intensity of selection value of 1.755, while retaining only 1% (i.e. a 10 fold reduction in selections) results in an intensity of selection value of only 1.52 times larger (i.e. $i = 2.665$).

Consider a simple example which is represented diagrammatically in Figure 7.1. Selection is to be carried out on a base population with an average, or mean, of 560 kg yield and with phenotypic standard

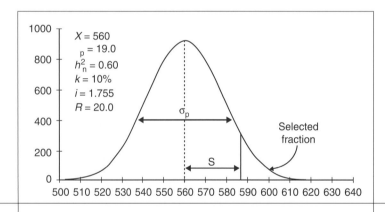

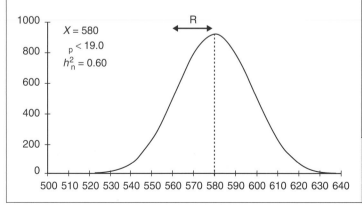

Figure 7.1 Illustration of the response from selection given population parameters from the unselected population (top) to predict the selected population (bottom).

deviation (σ_p) of 19.0 kg. If we assume that from past research it is known that the heritability (h^2) is equal to 0.6 and selection is to be carried out at the 10% level (i.e. $k = 0.1$, $i = 1.755$).

From this we have that:

$$\sigma i = 19.0 \times 1.755 = 33.34$$

From this we can estimate the performance of the selected fraction in the following year as the response to selection would be σih^2 and equal to $33.34 \times 0.6 = 20.0$ kg. The mean of the selected plants would therefore be 560 kg + 20 kg = 580 kg = the average performance of the top 10% selected lines in the next year.

It should be noted that the phenotypic standard deviation in the selected population must be less than the whole (unselected) population. As it can be assumed that the error variance remains constant, then this must mean that the genetic variance is smaller and the error variance is the same. From this, the heritability between the selected population and further selection years must be less than from the base population if the first selection year.

Therefore, if selection continues, then there would be decreasing response with increasing rounds of selection (Figure 7.2).

Return now to the response equation given above, and recall that the formula for the broad-sense and narrow-sense heritabilities is:

$$\sigma_g/\sigma_p \quad \text{and} \quad \sigma_a/\sigma_p, \text{respectively}$$

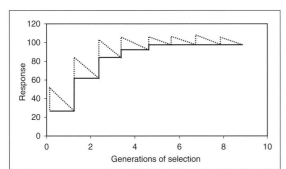

Figure 7.2 Response to selection from successive rounds of selection. The dashed line indicates the phenotypic expression and the solid line represents the genetic gain. Note that greatest gains are from the initial rounds of selection and that after several rounds of selection there is little or no gain.

where σ_g is the genetic variance component, σ_a is the additive genetic variance and σ_p is the phenotypic variance. From this we can write the average performance of a selected population after selection is:

$$P = X + \sigma ih^2$$

where X is the average performance of the initial population (i.e. the unselected family mean), i is the selection intensity, h^2 is the heritability and σ is the phenotypic standard deviation between plants in the population.

This means that the very best responses from selection are based on high family means, high selection intensity (although limited increase in return for very high selection), heritability and the phenotypic variance. From this breeders should be aiming to:

- Identify highly productive families with high average performance (i.e. high means)
- Maximize heritabilities by minimizing non-genetic errors. This can best be achieved by good experimentation, increasing plot sizes and replication levels
- Select as intensely as considered feasible, although remember the efficiency will only increase as a reciprocal beyond 20% selected
- Choose parents which are genetically diverse for characters that require improvement or change, and hence attempt to increase the phenotypic variance

On the other hand if a plant breeding programme is not producing the expected response, the same equation can be used to identify possible reasons for the failure.

The close correspondence between heritability and the proportional change in a selected character from one generation to the next when selection is applied has already been pointed out. Having considered estimation of narrow-sense heritability, h^2_n in some detail earlier, it is now appropriate to return to the issue of estimating heritability.

A third definition of narrow-sense heritability, usually termed the realized heritability, is:

$$h^2_n = R/S$$

where R is the *response to selection* (the same as described above) and S is the *selection differential*. The response to selection is the difference between the mean of the selected genotypes for a particular character and the mean of the population before selection was applied.

The selection differential is the average phenotypic superiority for the character in question of the selected genotypes over the whole of the population from which they were selected.

Consider the following example, the average seed yield of an F_3 family is 15 kg. Suppose plants that produced the highest seed yields (with a mean = 20 kg) were selected and grown to the next generation. What would be the selection differential? Since the selection differential is the average performance of the selected plants over the base population (i.e. all original unselected plants in the family) as a whole, $S = 20 - 15$ or 5 kg seed yield.

Now, if the mean seed yield of the selected progeny in the following year (F_4) was found to be 17.5 kg, what would be the response to selection? Since response to selection is the difference between the mean of the progeny and the mean of the parental generation before the application of selection, $R = 17.5 - 15$ or 2.5 kg of seed yield.

Finally the narrow-sense heritability, h_n^2 would be given by:

$$h_n^2 = R/S = 2.5/5.0 = 0.5$$

It should be noted that in the above example it is assumed that there are no dramatic year effects. In a practical situation, actual performance from year to year is highly variable. This can be taken into account in part by growing a *random sample* of progeny and controls the next year. Assuming that the random sample is indeed representative of the whole sample, it will be possible to use this in order to adjust the values and to obtain a direct indication of response to selection.

Similarly, a plant breeder is quite likely to want to know what response might be expected from a given selection differential when the narrow-sense heritability has already been estimated (from the partitioning phenotypic variances or from offspring-parent regression).

Therefore, if the selection differential (S) applied was 5 kg of seed yield and the narrow-sense heritability had been estimated to be 0.5, the response to selection expected would be:

$$R = h_n^2 \times S = 0.5 \times 5.0 = 2.5 \, kg$$

Thus the average seed yield of the selected progeny might be expected to be $15 + 2.5$ or 17.5 kg.

Association between variates or years

The degree of association between any two, or a number, of different characters can be examined statistically by the use of *correlation analysis*. As noted earlier, correlation analysis is similar in many ways to simple regression but in correlations both variables are expected to be subject to error variance, and there is no need to assign one set of values to be the *dependant variable* while the other is said to be the *independent variable*. Correlation coefficients (r) are calculated from the equation (see in more detail p. 76):

$$r = \frac{SP(x,y)}{\sqrt{[SS(x) \times SS(y)]}}$$

Diagrammatically, the association between two variables is shown in Figure 7.3 with positive correlation (top), no correlation (middle) and negative correlation (bottom).

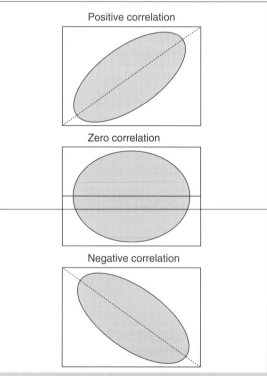

Figure 7.3 Diagrammatic representation of positive correlation (top), zero correlation (middle) and negative correlation (bottom).

As you can see, if there is high positive correlation between two variates it will be possible to select individual genotypes which have high expression in both traits. Conversely, if there is high negative correlation between variates it may be very difficult to select genotypes with high expression in both characters. The magnitude of the correlation value, in absolute terms, can be associated with underlying physiological processes or even pleiotropy (i.e. the same genes directly control expression in the two characters) or is a reflection of genetic linkage.

It would seem obvious that there must be some relationship between r, the correlation coefficient, and h^2, the heritability. If therefore characters are recorded on the same set of genotypes grown in two different environments (say locations or years), then the magnitude of the correlation coefficient indicates the relationship between performances in the different environments. Squaring the correlation coefficient (r^2) provides an estimate of the proportion of the total variation between the environments that is explained by the correlation. The total variation between sites can be considered the total phenotypic variation, and as the proportion accounted for by regression must have a genetic base, then a simple relationship exists, whereby r^2 is a direct estimate of h^2.

Heritability and its limitations

In this short but important section, a critical look is taken at the concept of heritability, its uses and misuses.

Four distinct methods of estimating narrow-sense heritabilities have been outlined:

- Partitioning of phenotypic variances
- Offspring–parent regression
- Response to selection
- Correlation

How response to selection can be predicted from a given selection differential when the narrow-sense heritability is already known from other experiments, has also been covered above. The concept of heritability, and estimates of it, have been of great value to plant breeders and to population geneticists interested in continuously varying characters in natural populations.

However, it is very important that the limitations of heritability estimates are realized. These limitations occur on at least three levels:

- There are many technical assumptions inherent in the theory as presented (e.g. that genes assort independently, that alleles segregate independently, that there is no epistasis). It is possible to allow for many of these complications, but only at the expense of making the theory more complicated
- An estimate of narrow-sense heritability strictly applies to a *particular* character, in a *particular* population, at a *particular* moment and in a *particular* environment. Thus, even for a single character, heritability is not constant. It is obvious that h^2 is particularly vulnerable to changes in environmental variance, changes that can occur at the same place at different times, different places at the same time, or both. But widely different estimates of h_n^2 for the same character can be found in different populations investigated at the same place and time. Also, it has been seen that additive genetic variance, and hence narrow-sense heritability, generally declines over generations of selection, even in a constant environment. Caution must therefore be exercised in interpreting estimates of h_n^2 if it is not known that every precaution has been taken to expose different populations and/or characters to the same range of environments and one is interested only in the response in the same, or very similar, environments
- While means are what are called first degree statistics, variances, etc. are second degree. Second degree statistics are usually 'less precise' than first degree. h_n^2 and h_b^2 (with the exception of h_n^2 from mid-parent onto offspring regression) being based on the ratios of variances, share all the weaknesses of second degree statistics.

Methods of selection

When a plant breeder is selecting a particular population for only a single trait the operation is usually relatively simple. The population is evaluated for the character in question and those phenotypes with desirable (whether this is high, low or intermediate) expression are *selected* while the phenotypes with less desirable expression are *rejected*. Therefore the only variable decision is the *selection intensity*, or the proportion of the total population that will be selected for further evaluation in relation to the proportion that are to be rejected or discarded. This form of selection is called *cull selection*. In such a scheme a target value is set and all phenotypes which meet the target are said to fulfill the selection criteria while those that do not reach the target value are rejected or fail to meet the selection criteria.

A successful new cultivar is rarely due to desirability in only a single character but is rather an overall increase over several different traits. Therefore deciding which individuals in a population are to be retained and which are to be discarded usually involves simultaneous evaluation for more than a single character.

When more than a single character is to be considered in a selection scheme a plant breeder can make selection by either *independent culling* of a number of characters or by using some defined *selection index*.

Independent cull selection

To examine this consider a simple case where there are only two variates to be included in the selection decision. If independent culling is used then the breeder will choose **target values** for each of the two characters independently. In order for a genotype to be selected, then the phenotype must exceed (or be less than, depending on the trait of interest) the target values of both of the characters simultaneously. Therefore each of the genotypes from the initial base population will fall into one of four possible categories. Which, for example if we are selecting for greater expression of both characters, will be:

- Greater than the target value set for both Variate 1 and Variate 2
- Greater than the target value set for Variate 1 but less than the target value set for Variate 2
- Less than the target value set for Variate 1 but greater than the target value set for Variate 2
- Less than the target value set for both Variate 1 and Variate 2

With this form of selection, only the genotypes which fall into category 1 (i.e. greater than both target values for each variate) would be retained, while all other categories would be discarded.

Index selection

Index selection involves creating an equation which includes values recorded for both variates. Selection indices can be either additive or multiplicative. For example, an additive selection index for the ith genotype with only two variates would be represented by:

$$I_i = (w_1 \times x_{i1}) + (w_2 \times x_{i2})$$

where I_i is the index value, w_1 and w_2 are the **weights** for each variate and x_1 and x_2 are the actual recorded values for each variate of the ith genotype. Obviously if **n** variates were included in the index value then the index equation would be represented by:

$$I_i = (w_1 \times x_{i1}) + (w_2 \times x_{i2}) + \cdots + (w_n \times x_{in})$$

A similar multiplicative index with only two variates would be:

$$I_i = (w_1 \times x_{i1}) \times (w_2 \times x_{i2})$$

where I_i, w_1, w_2, x_{i1} and x_{i2} are as above. Finally if **n** variates were included in a multiplicative selection index we would have:

$$I_i = (w_1 \times x_{i1}) \times (w_2 \times x_{i2}) \times \cdots \times (w_n \times x_{in})$$

The difference in results between index selection and independent culling are primarily related to the association between the two (or more) variates and the differences in the relative weighting of them. If there is good association between the variates (i.e. high expression in one variate is related to high expression in the other, and *vice versa*) and both are nearly equally valued, then there may be little difference between the genotypes selected by either method (Figure 7.4). If, however, there is poor association between variates (i.e. high expression in one trait is not related to a similar high expression in the other variate) or one character is of vital importance, then there could be a large difference in the genotypes that would be selected by index selection over independent culling (Figure 7.5).

In almost all studies carried out it has been shown that index selection is more effective in identifying genotypes that are 'superior' for many different traits. The difficulty in all selection index schemes is how to determine the *index weights* (i.e. the w_is).

It will not be possible within the scope of this book to fully explore the possibilities available with selection

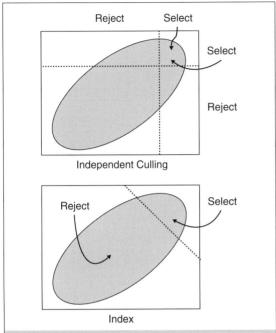

Figure 7.4 Association between independent culling and index selection when there is high correlation between two selectable traits.

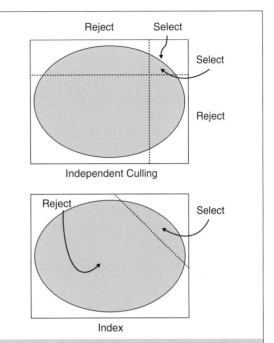

Figure 7.5 Association between independent culling and index selection when there is low correlation between two selectable traits.

indices. In simple terms, however, variate or character values in a selection index can be weighted either by:

- **Economics** where the potential economic impact of each trait is estimated and the datum recorded of each variate expression is weighted by that value. For example, the average price paid per unit weight can usually be predicted from past seasons and an increase in productivity could be related in money terms by an appropriate weight. Similarly if a particular insecticide costs a unit more per acre than if biological resistance is incorporated then that resistance will accrue monetary value. The problem with economic weights is that they change from year to year. If there is over-production of a product in any year then there is a tendency for the unit weight price to drop etc.
- **Statistical features** where the weight values are derived according to some statistical procedure. The most commonly used routines have involved *multivariate transformations* such as principal component analysis, canonical analysis or discriminant function analysis. Each of these statistical techniques,

although all called analysis, are in fact statistical transformations which produce various equations of multi-variate data, usually with minimum correlation between traits or maximum discrimination between genotypes. The problem with statistical weights is again that they will be different from data set to data set. In some cases the weights do indeed show some biological meaning but in other cases there appears to be no coherent association.

Selection indices can be extremely useful in plant breeding and their true value is perhaps yet to be realized. If index selection is carried out in a meaningful manner, then index selection should be more effective than independent culling.

Errors in selection

Each time selection is applied there is a chance that an error will occur. Errors in selection happen because the true genotype value is masked by environmental effects or because of administrative or clerical error.

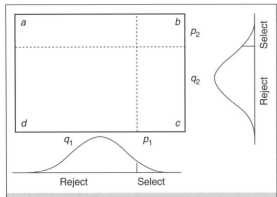

Figure 7.6 Classification of phenotypes based on independent culling of a single trait over two stages of selection.

Consider the illustration in Figure 7.6 which shows two stage selection of a single character. The distributions at the side and bottom of the figure show the frequency distribution of each stage. For simplicity assume that selection is being carried out over a two year cycle. In the first year, a proportion of the total genotypes (p_1) will be selected while the remainder (q_1) are theoretically discarded, where $p_1 + q_1 = 1$. Similarly, in year two a proportion of genotypes would be selected (p_2) and the remainder discarded (q_2), where $p_2 + q_2 = 1$. From this bi-variate distribution, each genotype is classified as:

a. rejected in the first year and selected in the second year
b. selected in the both years
c. selected in the first year but rejected in the second year
d. rejected in both years

From this there are two areas of misclassification and hence errors in selection. These have been termed:

Type I error where genotypes have been rejected in the first year and selected in the second. If the proportion of genotypes selected in year 1 is p_1, then the Type I error is calculated by $c/(c+b)$.

Type II error where genotypes are selected in the first year but rejected in the second year. If the proportion of genotypes selected in year 2 is p_2, then the Type II error is given by $a/(a+b)$.

In terms of practicality, Type I errors are far more important than Type II errors. Type I errors result from wrongfully rejecting a genotype, based on phenotypic performance in the early selection stage, which really should have been selected. Therefore this results in discarding potentially valuable genotypes. Type II errors result from selecting genotypes in the first year which really should have been discarded and therefore result in a waste of resources which should have been better used in other areas.

A second means to examine data from two stage selection frequencies uses **selection ratios**. Consider that all the population is evaluated in year 1 and selection is carried out such that a proportion of the population (p_1) is selected while the remainder (q_1) is discarded. Then all the population is re-evaluated in year 2 and again selection is carried out such that a proportion (p_2) is selected and the remainder (q_2) rejected. From this the question arises as to what proportion of the population selected in the second year would have been:

• Selected in the first year
• Discarded in the first year

The ratio of these two proportions is termed the **selection ratio** and is given by:

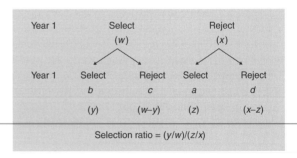

Values of selection ratios can range from 0 (zero) to infinity. If the selection ratio is equal to zero, then there was no repeated selection over the two stages. A selection ratio between zero and one indicates that a higher proportion of genotypes were selected in year 2 from those discarded in year 1 than were selected from year 1 (i.e. negative correlation). A selection ratio less than 1 therefore suggests that selection has had a negative effect (i.e. all the good genotypes have been discarded). A selection ratio equal to one indicates that selection has occurred at random (i.e. a zero correlation coefficient between the two stages). If the selection ratio is greater than one then selection has been better than random (i.e. a positive correlation between the stages). An increased magnitude of selection ratio shows increased efficiency of selection. For example if a selection ratio of 2.0 is obtained then the genotypes selected in year 1 would be twice as likely to be re-selected in year 2 than genotypes that were discarded in year 1.

It should be obvious that the selection ratio is related to the heritability of the character being selected. It also should be noted that the selection ratio will also be influenced by the selection intensity. Obviously, irrespective of the heritability, for a character, the selection ratio will be zero, if the selection intensity is set so high that no genotypes survive repeat selection. Similarly, the selection ratio will always be 1.0 where the selection intensity is so low that no genotypes are discarded. The relationship between selection ratio values and heritability is linear and related to selection intensity (Figure 7.7).

Similarly, it is possible to estimate selection ratio values for different selection intensities if the heritability is known. Where heritability is zero, then there is no response to selection and hence the selection ratio is zero. The selection ratios with different selection intensities and heritabilities are shown in Table 7.1.

Type I and II errors and selection ratios can be useful in setting the selection intensity levels at each stage in a breeding scheme. To estimate any of these it is necessary to determine the frequency of genotypes which fall into the *a*, *b*, *c* and *d* classes (Figure 7.6). In order to achieve this, it is necessary to artificially select and reject genotypes in one stage and to re-evaluate all selected and rejected lines in a second stage.

It has been mentioned above that correlation analysis can be useful in determining selection efficiency. This can be done by the use of *inverse tetrachoric correlation*. Tetrachoric correlations were first described by Digby in 1983. He showed that it was possible to determine the correlation coefficient between two stages of evaluation (i.e. two years of testing) from frequency tables like the one shown in Table 7.1.

Inverse tetrachoric correlations are indeed the inverse process where given the correlation coefficient between two selection stages it is possible to estimate the values of a, b, c and d (from Figure 7.6) and hence estimate Type I and Type II errors and selection ratios at different selection intensities.

The theory of tetrachoric correlations are beyond this book; however, values of the **b** (the frequency of genotypes that would be selected at both stages of a two stage selection) for varying selection intensities used at the different stages and with correlation values between the stages ranging from (0.2, 0.3, 0.4, 0.5, 0.6, 0.7, 0.8 and 0.9) are shown in Table 7.2.

To illustrate the use of this table consider the following example. It is known that the correlation coefficient for seed yield, between two assessment years (year 1 and year 2) is equal to 0.7. What would be the Type I error and Type II error given that selection was carried out at the 20% level in year 1 and at the 15% level in year 2. From the table with p_1 at 0.2 and p_2 at 0.15 and with a correlation coefficient of 0.7 we have a **b** value of 093 (or 93 genotypes out of 1000).

Table 7.1 Selection ratios values with different selection intensities and heritability values.

Selection intensity		Heritability			
Year-1 (%)	Year-2 (%)	0.2	0.4	0.6	0.8
5	5	6.98	13.95	20.93	27.91
5	10	3.04	6.07	9.11	12.14
5	15	1.82	3.63	5.45	7.28
5	20	1.22	2.45	3.68	4.91
10	10	4.02	8.03	12.05	16.07
10	15	2.27	4.55	6.83	9.11
10	20	1.50	3.01	4.51	6.02
15	15	2.88	5.76	8.65	11.53
15	20	1.87	3.74	5.60	7.47
20	20	2.33	4.67	7.00	9.34

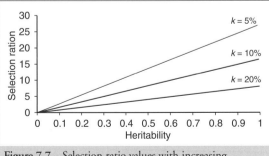

Figure 7.7 Selection ratio values with increasing heritability and different selection intensities.

Table 7.2　Number of genotypes (per 1000) that would be selected in both stages of a two-stage selection scheme (*b* values in Figure 7.6) based on inverse tetrachoric correlations from different selection intensity and correlation coefficients of $r = 0.1, 0.2, 0.3, 0.4, 0.5, 0.6, 0.7, 0.8$ and 0.9.

Year 1				
0.05	004 005 007 009 012 016 020 025 032			
0.10	007 009 012 016 019 024 029 035 043	013 017 022 027 032 039 047 056 069		
0.15	010 013 017 020 025 030 035 041 047	019 024 030 036 043 051 059 070 083	028 035 042 049 058 067 078 091 108	
0.20	013 017 020 025 029 034 039 044 049	025 031 037 044 052 060 069 079 091	037 044 052 061 070 081 093 107 125	048 057 066 076 087 099 113 129 150
	0.05	0.10	0.15	0.20
		Year 2		

From the fact that we selected 20% in year 1 and with 1000 genotypes this would have resulted in $0.2 \times 1000 = 200$ selected in total, then the number of genotypes selected in year 1 and discarded in year 2 (**c**, in Figure 7.6) would be $200 - 93 = 107$. Therefore that Type II error would be $107/200 = 0.535$, or 53.5% of genotypes selected in year 1 will be discarded in year 2. ~~Similarly, 15% ($p_2 = 0.15$) were selected in year 2.~~ So given that 1000 genotypes were screened, this would result in $1000 \times 0.15 = 150$ genotypes selected in year 2. From this we have that $150 - 93 = 57$ genotypes would have been selected in year 2 which would have been discarded in year 1. Therefore the Type II error is $57/150 = 0.380$ or 38% of all selections made in year 2 will have been discarded in year 1.

It can be seen therefore that even with relatively high correlations between different years of evaluation (i.e. $r = 0.7$) there will be a high potential selection error.

By subtraction we have that $1000 - (93 + 107 + 57) = \mathbf{d} = 743$, the number that would be discarded

in both stages. From this the selection ratio would be $[93/107]/[57/743] = 11.19$. Therefore a genotype selected in the first year would be more than 11 times more likely to be re-selected in a second year than a genotype discarded in the first year. Therefore despite the Type I and II errors, with a correlation between stages of $r = 0.7$ then selection is more than effective at these selection intensities.

It is interesting to try the same operation with:

- Uniform selection intensity and varying correlation coefficients
- Uniform correlation coefficient and varying selection intensity

Inverse tetrachoric correlations can also be used in a similar way to determine the association between selection for two different characters if the correlation coefficient between the traits is known.

APPLIED SELECTION

Selection in a plant breeding programme is an operation which is carried out over several years. After genetic variation is created, then a population of genotypes will be evaluated under different environmental conditions. At each stage the '*most desirable*' lines are selected, while the lines with defects or that are less desirable are discarded.

For simplicity the various stages of selection can be divided into three types:

- **Early generation selection.** This is the first stage where the initial unselected population is screened. In most programmes this critical stage can involve many thousands of individuals (Figure 7.8).
- **Intermediate generation selection.** After the least adapted lines have been discarded more detailed evaluation of lines is carried out. Intermediate selection usually involves hundreds, rather than thousands of lines.
- **Advanced generation selection.** At this stage the initial population has been reduced such that only tens of lines have survived. Trials in the advanced selection stage are most accurate and detailed, including multiple location evaluation trials.

Figure 7.8 Selecting single potato plants based on visual appearance.

Number of genotypes in initial populations

To many, plant breeding is a *'numbers game'* where the more genotypes screened results in a greater chance of identifying desirable recombinants which will eventually become new cultivars.

Why therefore is it necessary to evaluate so many different genetic lines in a plant breeding programme? Consider a simple wheat breeding scheme which has the objective of developing new cultivars that have high yield, good bread making quality, qualitative resistance (dominant) to stripe rust, quantitative resistance to mildew, good establishment, cold tolerant, short in stature, early maturity and resistant to lodging. Consider that the selection intensities in Table 7.3 are needed to ensure that at least some individuals will meet the required standards. Also shown is the accumulative frequency of desirable individuals, given that all characters are independently inherited.

Obviously, even when this very limited set of criteria is used as the Objective, and with relatively modest selection intensities, the number of plants that need to be screened can be very high. If a breeder wishes to have some chance of success in having *at least one*

Table 7.3 Possible selection intensities used in a wheat breeding scheme to develop cultivars that have high yield, good bread making quality, qualitative resistance to stripe rust, quantitative resistance to mildew, good establishment, cold tolerant, short in stature, early maturity and resistant to lodging. Also shown is the accumulative frequency of desirable genotypes, given that all characters are independently inherited.

Character	Selection intensity (%)	Accumulation frequency
High yield	5	1 : 20
Good quality	5	1 : 400
Stripe rust resistance	50	1 : 800
Mildew resistance	20	1 : 4000
Crop establishment	50	1 : 8000
Cold tolerance	20	1 : 40 000
Short stature	10	1 : 400 000
Early maturity	10	1 : 4 000 000
Lodging resistance	10	1 : 40 000 000

individual that meets the criteria the numbers that need to be screened will be large.

Plant breeding therefore does require evaluation of many thousands (or millions) of plants to have any chance of producing a successful new cultivar. So given

Table 7.4 Selection scheme used to develop new potato cultivars at the Scottish Crop Research Institute.

Year	Number of genotypes	Number of replicates	Plot Size	Characters assessed
1	140 000	1	1	Visual assessment of commercial worth
2	40 000	1	1	Visual assessment of commercial worth
3	4000	1	3	Visual assessment of tuber size, shape, tuber number, yield and defects
4	1000	2	5	Actual assessment of yield, initial quality tests, visual assessment of appearance and defects
5	500	2	10	Actual assessment of yield, fry quality, boil quality, initial disease testing for late blight and common scab, visual assessment of appearance and defects
6	100	2	40	Yield, fry and boil quality assessment from early and late harvest, initial taste testing, multiple disease testing, initial virus testing, visual assessment of appearance and defects
7	50	4	40	Yield, quality and disease testing at seven locations throughout the target region
8	10	4	40	Repeat multiple locations testing, initial on-farm testing (large field scale trials)

that the initial population needs to be large and that selection of the better lines should be carried out as efficiently as possible, how can the best genotype be identified?

To further examine the different stages of selection consider the two examples. The selection scheme used to develop new potato cultivars at the Scottish Crop Research Institute is shown in Table 7.4.

In this scheme years 1, 2 and 3 would be considered to be early generation selection, years 4, 5 and 6 would be intermediate generation selection and years 7 and 8 would be advanced selection.

The wheat breeding programme at the University of Idaho has the selection scheme shown in Table 7.5.

In this scheme years 1 and 2 would be considered as early generation selections, years 3 and 4 would be intermediate selection and years 5 and 6 would be advanced generation selection. In year 7, and subsequent years, remaining selections (2 to 3 lines) would be entered for regional testing where they would be evaluated at many western USA locations.

Therefore early generation selection should eliminate the very worst genotypes, intermediate selection would identify the very best genotypes and advanced selection would confirm genotypic performance over differing locations, assess environmental stability and identify the

superior (cultivar quality) genotypes from those which may just fail to become cultivars.

After each stage of selection fewer genotypes will remain for further testing. Increased rounds of selection will also be associated with decreased genetic variation between selected lines. It will therefore require more detailed evaluation studies to differentiate between remaining selections.

Early generation selection

Selection in the early generation stages differs from later selection because:

- Many thousands of lines are to be screened
- Only small amounts of planting material are available from each genotype and so sophisticated experimental designs with large numbers of plots and high replication are not possible
- Selection is often carried out on highly heterozygous populations where dominance effects can be large and can mask the true genotype being selected

The first two points can be considered as a single problem because, even in cases where large quantities

Table 7.5 Selection scheme used by Dr. Robert Zemetra in the soft white wheat breeding program at the University of Idaho.

Year	Number of genotypes	Plot type	Characters assessed
1 – F$_3$	1 200 000	Single plants	Visual selection of plant types, plant height and stripe rust resistance
2 – F$_4$	20 000	Head rows	Visual selection of plant types, uniformity, yield, height, stripe rust resistance, lodging resistance. Actual assessment of protein content and kernel hardness
3 – F$_5$	500	Preliminary yield trials, at one location	Actual yield performance, stripe rust yield resistance, lodging, stand establishment, heading date, test weight, dough viscosity, milling quality, baking quality, protein content and kernel hardness
4 – F$_6$	80	Preliminary yield trials at two locations	Actual yield performance, stripe rust yield resistance, lodging stand establishment, heading date, test weight, dough viscosity, milling quality, baking quality, protein content and kernel hardness, yield stability, Russian wheat aphid resistance, dwarf bunt, cephalosporium stripe
5 – F$_7$	10	Advanced yield trials at eight locations	Actual yield performance, stripe rust yield trials, resistance, lodging stand, establishment, heading date, test locations weight, dough viscosity, milling quality, baking quality, protein content and kernel hardness, yield stability, Russian wheat aphid resistance, dwarf bunt, cephalosporium stripe, foot rot, Hessian fly resistance and winter-hardiness
6 – F$_8$	6	Advanced yield trials at eight locations	Actual yield performance, stripe rust resistance, lodging, stand establishment, heading date, test Weight, dough viscosity, milling quality, baking quality, protein content and kernel hardness, yield stability, Russian wheat aphid resistance, dwarf bunt, cephalosporium stripe, foot rot, Hessian fly resistance and winter-hardiness

of planting material are available, it may be impractical to have very large plots and high replication of so many different lines. Staff and land are not usually available to carry out such large screens.

Early generation selection is therefore carried out on small plots and most often on un-replicated plots. Even when the test entries are not replicated it is possible to increase the efficiency of testing by including a wide range of control entries inter-spaced within the test plots. Direct comparison can be made between the control lines (often existing cultivars) and those under test. The control plots can be included more than once over the whole trial area and from this an estimate of plot to plot error variance can be obtained.

Several forms of analysis are available by which test entries can be compared to, or adjusted to, adjacent control lines. It should be noted that it is unlikely that all controls within a trial will have equal performance characters. Some controls will have higher yield, others would have better disease resistance or high expression

in a single character, and this needs to be accounted for if comparisons are to be made with test lines.

Visual assessment

The large number of genotypes which need to be assessed in the early generations usually dictates that most selection is by visual assessment. Visual assessment of genotype performance is based on a mental image of the desirable attributes (*ideotype*) that will constitute a successful variety. Such assessments are therefore similar to an informal selection index. The efficiency of visual assessment can be influenced by breeders' experience and also the time taken over assessment. Visual assessment has been proven to be more effective when more than a single breeder is involved in assessment and selection is carried out based on the average assessment rating.

Throughout the season different traits can be visually assessed and then genotypes culled based on this information. At harvest, only the lines which have met all selection criteria are retained and others are discarded.

Visual assessment is often very subjective and different people have been shown to give differing emphasis in screening, depending on individual preference. Overall, however, different evaluations based on visual ratings can be remarkably similar in the lines that are chosen. Despite the problems with visual selection it can be carried out relatively quickly so many lines can be evaluated and at low cost. The biggest failing of visual selection alone is that characters can only be assessed if they can be seen. Therefore it is not usually possible to use visual evaluation to screen, for example, for quality characters. If more objective selection is to be achieved than it may be necessary to actually record information on yield, disease rating or quality.

Even when there is no replication it is possible to obtain some indication of error in visual assessment if the assessment operation is repeated. This will not provide environmental error estimation but can often be useful in determining the repeatability of visually assessed characters.

Mass selection

It is often possible to use mass selection in the early generations. Examples of this would include selection for short plants by cutting tall ears from populations in wheat. Mass selection can also be used to select larger seeds, higher specific gravity in tubers and morphological traits such as fruit colour.

Mass selection in the early generations can be achieved by growing the early population under specific environmental conditions. The more adapted lines will be more productive and the frequency of less adapted genotypes will reduce. Bulk selection has been shown to be effective in increasing the frequency of drought, heat, salt and other stress tolerances.

Efficiency of early generation selection

The efficiency of early generation selection has been examined in a number of different crops. When breeding an autogamous species, for example wheat or barley, selection will be influenced by the highly heterozygous nature of the breeding lines in the early generations. Segregation effects can be avoided by advancing towards homozygosity prior to selection but has not been common in the past because of time restraints or cost factors.

Visual assessment of yield and yield components has been examined and visual evaluation of yield from single rows or small plots has proved unreliable in predicting actual yield in subsequent generations. The highest yielding progeny bulks, derived from F_2 and F_3 single plants, do not necessarily produce the highest yielding segregants. Visual selection for yield on individual plants in cereals results in only a random reduction in population size with little or no effect in increasing yield. Even when the actual yield of an early generation of a cereal pedigree bulk breeding scheme (say F_2 or F_3) was measured and it was found to be significantly correlated with yield in later generations (say F_5 or F_6) the association found between segregating populations was usually so poor that it was questionable whether selection at the early stages (along with the expense that this would incur) would be justified.

Selection for yield *per se* in the early segregating generations of other inbreeding species has also been shown to produce an effect which is no better than random. Examples from past research include chickpea, cotton, soybean and rice. In addition selection for yield components such as seed size in chickpea and grains per ear in spring barley was shown to be slightly more effective in the early generations than selection for yield itself.

The large numbers and small plots used in the early generations dictates that selection is only carried out for characters which are highly heritable. Often these only include single gene traits. As might be expected, selection for qualitative disease resistance in the early generations has been found to be more effective than selection for quantitatively inherited resistance or other polygenic characters.

The efficiency of selection in a pedigree bulk breeding scheme has been related to the heterozygosity of the bulks under selection. As homozygosity increases, selection becomes more effective. Homozygosity can be accelerated by single seed descent. However, care is needed to ensure that in single seed descent there is not a non-random loss of genetic material. Homozygosity can also be accelerated by various doubled haploid techniques. Again however, care must be taken in the use of these procedures as there is evidence of non-random success and a strong genotypic response to *in vitro* regeneration.

Most research into the efficiency of selection on clonally reproduced crops has been on potatoes and sugarcane. Early generation selection in potatoes, see Figure 7.8 has been shown to be at best a random reduction in the number of genotypes in the breeding scheme. There has been some evidence that selection in the early generations was producing an undesirable response and that selections were not always the genotypic lines most suited to agricultural conditions.

Significant correlations have been found between sugar cane seedlings and later clonal generations for stalk and stalk diameter. These associations although statistically significant would result in large selection Type I errors and selection ratios less than 2.0. With such results it may be difficult to justify the expense and effort that such selection would involve.

Early generation selection of grasses using small plots resulted in identifying lines which did not perform well under sward conditions where inter-plant competition was greater. A similar response has been noted in potato where selected lines were less competitive under field stand conditions due to selection being carried out under wide plant spacing.

Despite the relative inefficiency of selection in the early generation stages most plant breeding schemes usually discard by far the greatest proportion of genetic variation in the first and second rounds of selection. It is certainly not uncommon to have cases where 99.9% of genotypes are discarded in the first or second selection stage and that this has been achieved using small plots and without replication.

In summary, selection in the early generations is usually affected by:

- **Limited amounts of planting material** so it is not possible to have sophisticated experiments involving large plots, high replication and multiple sites
- **Large numbers of genotypes** need to be evaluated which also usually results in small plots (often single plants), low levels of replication and single location trials

As a result, many of the initial evaluations are carried out by visual inspection rather than, say, actual recording yield. Similarly many of the more "difficult to assess" traits including polygenic disease or pest resistance or quality character cannot be easily taken into account.

It is difficult to determine exactly what characters are to have priority in early generation selection. Unfortunately there is not any simple equation which

allows breeders to say that it is best to select for this now and at this intensity. There are some simple questions which can help in making these decisions. These include:

- What are my breeding objectives and what characters are to be included throughout the **whole selection process**?
- What characters can be most easily and most economically assessed on small plots with minimal replication?
- Which characters are most heritable (i.e. high h^2)? Which have low heritability? Selection in the early generation should be based on the most heritable traits
- Which characters have highest priority? For example which characters **must** a new cultivar have? Either by being important (i.e. high yield or specific quality) or by legislation (i.e. low glucosinolates and erucic acid content in canola)

Never forget the golden rule of any selection, that a breeding scheme should never carry more individual genotypes than can be **efficiently screened**. It is almost always more effective to evaluate fewer lines with greater accuracy than to use an ineffective selection scheme.

Intermediate generation selection

It is assumed that at the intermediate generation selection stage, the large initial population (usually thousands) has been reduced to a practical number which will allow more detailed assessment (usually hundreds of lines). It is also assumed that by this stage there has been a simultaneous increase in the availability of planting material. As a result of fewer lines and more planting material, it is possible to organize evaluation trials which have reasonable plot sizes (may differ according to the crop), and replication of all test entries is possible.

The number of lines that require testing at the intermediate stage will still be large enough to dictate that evaluation is still restricted to only one (sometimes two) locations.

Field trials
Field testing is a major part of all selection, and intermediate selection is no exception. Test entries should

always be evaluated in comparison to control entries in replicated yield trials. Randomized Complete Block (RCB) designs are commonly used for the first rounds of intermediate generation selection. These designs provide reasonable error estimates and are fairly robust. One major advantage of RCB designs is that they can be used for any number of entries. As the number of surviving test entries is reduced then more detailed incomplete block designs such as lattice squares, rectangular lattices and partially balanced incomplete block designs may be used.

Lattice squares are amongst the most efficient designs that can be used for field testing in a plant breeding programme. A lattice design is similar to a RCB in that each entry appears once in all replicate blocks. However, within each replicate block, plots are arranged into sub-blocks. Analysis of data from lattice designs allows the actual mean performance of each test entry to be adjusted due to two dimensional (row and column) environmental variation. The major problem with lattice squares is that the number of test entries must be an exact square (i.e. 9, 16, 25, 36, 49, 64, 81 and 100, etc.). A second restriction is that the most efficient use of the designs requires high replication. For example a 16 entry design (4×4 lattice square) requires $4 + 1 = 5$ replicates. Larger lattice squares can be used with $n - 1$ replicates, where n^2 is the number of entries in the whole trial.

Rectangular lattice designs allow greater flexibility in the number of entries and replicates used, although each replicate must be a rectangle (i.e. 10 plots $\times$ 5 plots, where sub-blocks would be either 5 to 10 plots). Rectangular lattice designs are not as efficient as lattice squares in reducing error variance as sub-block adjustments are made in only one direction.

One advantage of all lattice designs is that they are resolvable (i.e. data collected from them can be analyzed as a RCB design).

Spilt-plot designs are often used in the latter stages of intermediate selection. The main use is often to evaluate a number of lines under differing environmental conditions at a single location. For example, several test entries may be assessed under differing nitrogen levels where genotypes would be main-plots and varying nitrogen levels are sub-plots. Similarly, spilt-plot designs can be used for differential chemical treatments or harvesting dates.

Variates recorded

Whereas in the early generation stages there are many thousands of test lines, there are very few characters recorded on each line. In intermediate selection the number of traits on which selection is based is increased, often considerably.

Data will be collected prior to planting, throughout the growing season, at harvest and post-harvest. A major part of plant breeding is managing the vast data sets which can arise and to interpret this information to best advantage in selection.

Data analysis and interpretation

It is useful to analyze data as they are collected throughout the year, so as not have a backlog of analysis which is needed for decision making at the end of the season. It is common in plant breeding to have a relatively quick turn over. For example in winter wheat breeding, evaluation plots are harvested, yields recorded, samples taken for quality assessment, assessment carried out and decisions made within a few months so that selection procedures can effectively use all possible information while still being able to plant selected lines at the appropriate seasonal time.

If selection is to be successfully applied for any character it is important that there are indeed significant differences between test entries. Obviously if an analysis of variance shows no significant difference between test lines for yield, then there is no genetic variation for the character, and hence there will be no response to selection.

It can often be useful to estimate narrow or broadsense heritabilities from yield trials. Broad-sense heritabilities can be easily obtained by simply estimating the genetic component of variance. In cases where test lines are highly heterozygous then error variances can be estimated from homozygous control entries in trials. Narrow-sense heritabilities can be estimated by regression if a sufficient number of the parental lines are included in the evaluation trial.

Heritabilities can also be estimated in relation to response to selection. To achieve this with any accuracy it is necessary to retain a certain proportion of the unselected population and to include these *random* selections along with the deliberate selections in the following seasons' trial. The practice of retaining a random sample is highly recommended as it allows continual check of what advances selection is making in producing more desirable lines.

Bar charts or histograms of data can be helpful in understanding the variation and distribution of data for individual traits. Inspection of distributions along with trait means and variances can help to determine possible culling levels (i.e. target values which must be met to be retained).

After each character has been analysed individually, it is very important to consider the relationships which exist between different traits. This can be achieved by simple correlation analysis.

If two characters of interest are positively correlated then there may not be any difficulty in selection (except that there will tend to be greater emphasis in either independent culling or index selection with positively correlated traits). However, if the expression of two characters is negatively correlated it may be impossible to select for high expression in both traits simultaneously. Lower culling levels or index selection will be necessary to identify lines which may be intermediate performance for both characters, as the more desirable recombinants are not present in the sample of materials evaluated.

Correlation analysis can be easily carried out using a variety of different computer software packages. The use of computers in all aspects of plant breeding will be discussed later. It is sufficient to say at this time that selection is one area where statistical analysis is helpful in understanding the vast data sets which are likely to arise and can also act as useful tools to select the better lines based on the data collected.

Selection

Most effective selection will result from most accurate data collection and highest heritability. This in turn is related to good experimentation.

A good understanding of the relationship between different characters and genetic variation within characters can be of tremendous help in deciding whether to apply independent culling (along with the cull levels) or whether a selection index is more appropriate.

Advanced selection

At the advanced selection stage it is assumed that all remaining genotypes in the selection scheme have previously been assessed for all (or the majority of) characters of interest to the breeding objectives. At this stage there are relatively few (under 100, and no more usually than between 10 and 50 lines) selections that have survived the previous selection stages. Selections would therefore be expected to have shown some value for yield, quality, disease resistance, pest resistance and be relatively free from obvious defects.

At this stage it is also assumed that there are relatively large amounts of planting material which allows evaluation at a number of different locations throughout the target region.

The major aims of advanced selection are:

- To confirm the past performance of selected lines over a wide range of different environments (locations and years)
- Identify the superior lines based on either specific or general adaptability

There is usually little response to selection at the advanced stages because most of the genetic variation has been reduced; the few remaining genotypes represent a highly selected (see response to selection in Figure 7.2) group.

Choice of advanced trial locations

The choice of land suitable for trials is discussed in the field plot techniques section. It is sufficient to emphasise some points and state a few additional factors here.

Locations for advanced testing must be representative of the environments in which the potential new cultivars will be grown. If the target region has several diverse locations (i.e. large differences in rainfall, temperature, soil type), or if different agronomic practices are applied in different regions (i.e. irrigation versus rain fed) then attempts should be made to ensure that at lease one location is chosen to represent each environment type.

If trial sites are a long way from the central research offices then it may be necessary to find good collaborators (farmers, county agricultural agents, extension personnel) who will take care of the research plots.

In some instances trials may be grown in other states or different countries. In this case it may be necessary to arrange appropriate phytosanitary inspections of the previous seed crop or the seed that will be planted.

Figure 7.9   Breeder's trial of advanced selections planted within a farmer's field crop.

Number of locations

The number of locations used for testing will be dependant on:

- Availability of planting material. This is not usually a major restraint at the advanced selection stage but may need to be considered.
- The diversity of environmental conditions throughout the target region. Obviously, if for example, the target region is the whole of the United States, then many sites will be needed, while if a small county is the only target area of interest then perhaps one or two local trials will be sufficient.
- The magnitude of error variances and genetic variances applicable to specific trials in any one year or location. If, from past experience, the environments for which a new cultivar are targeted are all very similar (i.e. small variance between sites) then fewer locations need be considered. If different regions are markedly different, then more sites would be required.
- The cost of individual trials and the availability of sufficient funds to pay for off-station trials. In the real world, most breeding programmes have restricted budgets and multi-location trials can often be expensive in shipping, land rental, staff time and travel.

There have been many debates regarding the substitution of more locations in advanced trialling at the expense of reducing the number of testing years. Obviously if more locations are evaluated in each year then fewer years may be necessary to fully evaluate environmental response of test lines. However, it should be noted that year to year environmental variation is almost always unpredictable (i.e. climatic), while with between location variation there will be many predictable environmental effects such as soil type. Therefore, it will always be necessary to assess advanced breeding lines over more than a single season, and several years testing may be required before a satisfactory decision of commercial worth can be made. Also, in practice, when appropriate data have been analyzed the interactions with years are often larger than those with site or location.

It should be noted that selection is still being carried out amongst lines in the advanced breeder trials and that this can affect the average performance over years. For example, consider that in the first advanced trial 50 breeding lines are tested. The *best* 25 lines will be re-evaluated in year 2 based on their *phenotypic* performance in the year 1 trial. Say, the *best* 5 lines, now based on phenotypic year 2 performances, are retained from the year 3 trial and tested in the third year. After the third year trial, the *best* breeding line will be considered for cultivar release. At this stage, and before, it is common practice to examine the performance averaged over 3–4 years of advanced testing and to compare this performance to standard control cultivars that were included in the trials. The performance of the breeding lines is likely to be somewhat biased as they had specifically been chosen in the previous years testing because they *had better than average* performance. In order to get a true representation of the new cultivars worth, it is common to evaluate the newly released cultivar for several years in breeder trials after selection is complete. These post selection trials are commonly conducted on a large scale, large (on-farm) plots and utilizing farm-scale equipment (planters, harvesters, etc.). Often it is the produce from these on-farm tests that offers breeders the first opportunity to have sufficient volume of material for actual quality evaluations.

Experimental design

Limited number of test entries combined with large amounts of planting material allows the use of the most sophisticated experimental design at the advanced stages. Therefore it is common to use lattice squares or rectangular lattice designs for location trials. It should be noted that such trials are often managed by collaborators who are inexperienced in handling trials and the more highly sophisticated the designs the more easily can it be planted or harvested incorrectly. With this in mind, randomized complete block designs may still offer the most practical design for advanced trials.

Genotype by environment interactions

A major goal of advanced selection is to determine the response of selected lines over differing environments. It is therefore difficult to consider this aim without specifically considering genotype by environment interactions.

As noted earlier, genotype × environment (G × E) interactions occur because some genotypes perform to a high degree under some environmental conditions while others perform poorly in that same environment, conversely the lower yielding lines may exceed the higher yielding genotypes when grown under different conditions.

G × E interactions affect traits throughout all stages of a plant breeding selection programme, unless molecular markers or other similar techniques are used to evaluate genotypes free from environmental influences.

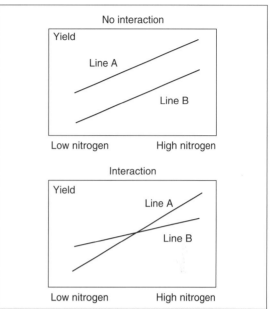

Figure 7.10   Genotype by environment relationships: where there are no genotype by environment interactions (top) and where the two genotypes (A and B) respond differently to different environments (nitrogen levels).

However, it is not usually possible to fully evaluate the G × E phenomenon until the advanced stages.

G × E interaction based on only two genotypes is shown diagrammatically in Figure 7.10. It should be noted that a significant interaction in an analysis of variance can be obtained even when there is no change in ranking of genotypes under study (no cross overs in performance). Consider the two genotype cases in Figure 7.10. On top both genotypes perform relatively similarly (i.e. the two response lines are parallel). Interactions can, however, appear significant if the lines converge, but do not cross (i.e. no change in ranking of the genotypes). If there is no change of ranking, the interaction can usually be designated as a **scalar effect** and is not considered by many as a true interaction. Some would argue that the lines would converge and eventually cross either above or below the range of the data set collected, but there is no evidence that this is true. Interactions caused by scalar effects can often be removed by transformation of the data. True interactions, where there are changes in ranking of lines in differing environments, cannot be eliminated by data transformation.

Environments can be classified into two different forms:

- Semi-controlled environments, where they are controlled by the grower or where there maybe little change over years or seasons. For example, soil type, seeding density, fertilizer application
- Uncontrolled environments, where there is often no chance of predicting conditions from one year to another. For example, rainfall, temperature, high winds

Obviously, even with the uncontrolled environmental conditions there can be some degree of prediction. For example, there will always be very low temperature in North Dakota in January and February while it will tend to be continually warm in Death Valley, California, during the summer months.

The early and intermediate selection have been carried out at a single (or few) location and so any surviving line should have at least been tested over more than a single year (albeit at a common location). It is difficult to carry out actual G × E studies on data where a large proportion of lines have been selected according to that data. For example if there are three years of data available from an intermediate selection stage which begins with 1000 lines and reduces these to 20 lines based on phenotypic data collected over the year then all the remaining genotypes will (by definition be those that were selected) have high phenotypic expression in all years.

Multiple location trials are therefore necessary for two reasons:

- To identify particular genotypes which perform well over a wide range of environmental conditions. These lines are said to have **general adaptability**
- To identify genotypes which perform to a high degree at specific locations or under particular conditions. These lines are said to have **specific adaptability**

Analysis of location trials

Various methods have been proposed for the statistical analysis of interactions in general and G × E interactions in particular. The existence of interactions between genotypes and environments was recognized by Fisher and Mackenzie, even before the formation of the analysis of variance.

There are several conditions that are assumed when carrying out an analysis of variance (and indeed also some other analyses). These include: randomness; normality; additivity and homogeneity of error variance. The latter of these is the one which usually causes most concern when carrying out analysis of variance of multiple location trials.

If there are only two experiments (i.e. two locations, two years, etc.) then the plant breeder can simply perform an analysis of variance on each experiment separately and from each obtain an estimate of the error mean square (σ^2). The larger of the two σ^2s can be used as the numerator and the smaller the denominator to carry out an F-test with the appropriate degrees of freedom. The resulting statistic can be compared to expected values from tables to determine whether the two σ^2s values are indeed different.

In cases where more than one experiment is being considered (i.e. an experiment carried out over three different years), a different approach needs to be considered. The F-test can still, however, offer a simple test where the largest σ^2 from the experiments is compared by dividing by the smallest σ^2 value. However, a more accurate method is available called a *Bartlett Test*.

Bartlett test

There are two forms of the Bartlett test depending on whether the variances (σ^2) to be compared all have the same number of degrees of freedom, or whether they have different degrees of freedom. The first of these two situations will be considered first.

When all σ^2 values are based on the same degrees of freedom the Bartlett test takes the form:

$$M = \mathrm{df}\{n\ln(\underline{S}) - \Sigma\ln\sigma^2\}$$

where $\underline{S}$ is:

$$\underline{S} = \Sigma\sigma^2/n$$
$$C = 1 + (n+1)/(3n\mathrm{df})$$

where n is the number of variances to be tested, df is the degrees of freedom that **all** variances are based on, ln refers to natural logs.

Now the Bartlett test in this case reduces to a chi-square (χ^2) test with $n - 1$ degrees of freedom, where:

$$\chi^2_{n-1} = M/C$$

Consider this simple example involving four variances (σ^2) all based on 5 degrees of freedom:

d.f.	σ^2	$\ln\sigma^2$
5	178	5.182
5	60	4.094
5	98	4.585
5	68	4.202
Total	404	18.081

$\underline{S} = 100.9$; $\ln(\underline{S}) = 4.614$

$M = (5)[(4)(4.614) - 18.081] = 1.88$, with 3d.f.

$C = 1 + (5)/[(3)(4)(5)] = 1.083$

$\chi^2_{3df} = 1.88/1.083 = 1.74$ns

As the χ^2 value in this case is smaller that the corresponding value from tables with 3 degrees of freedom, we would say that the four variances were *not significantly different* at the 5% level.

Consider now the case where each of the variances under test is based on different degrees of freedom. Now the Bartlett test is based on:

$$M \quad (\Sigma\mathrm{df})\ln(\underline{S}) - \Sigma\mathrm{df}\ln\sigma^2$$

where $\underline{S}$ is equal to:

$$\underline{S} \quad \Sigma\mathrm{df}[\Sigma\mathrm{df}.\sigma^2]/(\Sigma\mathrm{df})$$

and

$$C1 + \{(1)/[3(n-1)]\}.[\Sigma(1/\mathrm{df}_i) - 1/(\Sigma\mathrm{df})]$$

and

$$\chi^2_{n-1} = M/C$$

To further examine this consider the simple example involving five variances, each based on a different number of degrees of freedom:

df	σ^2	$\ln\sigma^2$	$1/\mathrm{df}$
9	0.909	−0.095	0.1111
7	0.497	−0.699	0.1429
9	0.076	−2.577	0.1111
7	0.103	−2.273	0.1429
5	0.146	−1.942	0.2000
Total 37			0.7080

$$\underline{S} = \Sigma\mathrm{df}\sigma^2/\Sigma\mathrm{df} = 13.79/37 = 0.3727$$
$$(\Sigma\mathrm{df})\ln(\underline{S}) = (37)(-0.9870) = -36.519$$
$$M = (\Sigma\mathrm{df})\ln(\underline{S}) - \Sigma\mathrm{df}\ln\sigma^2$$
$$= -36.519 - (-54.472) = 17.96$$
$$C = 1 + [1/(3)(4)](0.7080 - 0.0270)$$
$$= 1.057$$
$$\chi^2_{4df} = M/C = 17.96/1.057$$
$$= 16.99^{***}, \text{ with 3d.f.}$$

As this value of chi-square exceed the value from statistical tables with three degrees of freedom we can say that the five error variances show significant ($p < 0.01$) **heterogeneity**.

It should be noted that the Bartlett test is *over* sensitive to deviations and it is usual under practical situations to only consider heterogeneity of variance occurring where we have significant at the 99.9% level, or higher.

Detecting significant treatment × environment interactions

The most common method, by far, of detecting genotype × environment interactions is to carry out an appropriate analysis of variance.

Various methods have been proposed for the statistical analysis of interactions in general and genotype × environment interactions in particular. Consider first a simple example where a number of genotypes are each grown in a number of different environments. Variance components can be used to separate the effects of genotypes, environments and their interaction by equating the observed mean squares in the analysis of variance to their expectations in the random model. Therefore, in terms of a mathematical model the yield y_{ijk} of the kth replicate of the ith genotype, grown in the jth environment can be estimated by the formula:

$$y_{ijk} = \mu + g_i + e_j + ge_{ij} + E_{ijk} \qquad (7.1)$$

where μ is the average yield over all trials, g_i is the effect of the ith genotype, e_j is the effect of the jth environment, ge_{ij} is the interaction between the ith genotype and the jth environment and E_{ijk} is an error term. Also:

$$\Sigma_i g_i = \Sigma_j e_j = \Sigma_{ij} ge_{ij} = 0$$

The expected mean squares from an analysis of variance are dependent on whether main effects are fixed or random. In the case of genotype × environment trials it is usually assumed that both effects are random. From this, given a set of trials with g genotype entries, r replicates at each location and l locations the following expected mean squares would be obtained from an analysis of variance, shown in Table 7.6.

In this case the genotype × location interaction would be tested for significance against the pooled replicate w locations × genotypes interaction. The effect of genotypes would be tested in an F-test against the interaction G × E. Replicates w locations would be tested against the replicate error and the effect of locations would be tested against the replicates w locations mean square.

When both years and locations are included, often different locations (or different numbers of locations) are used in each year. In these cases the analysis reduces to genotypes × environments interactions where environments include a combination of years and sites (i.e. y years and s sites would result in ys environments). Environmental differences are often greater over years than over locations and it can be informative to separate year and location effects in the analysis of variance.

More complex models and expected mean squares can be derived for location trials grown over more than one year which produce year × genotype, location × genotype and year × location × year interactions. In this case the expected mean squares (EMS) would be those shown in Table 7.7.

Given that a significant interaction is detected in the analysis of variance, and this is not due to a scalar effect, then it can often be useful to have additional information regarding the partition of the interaction variance.

Worked example

Consider the example where 20 spring canola (*B. napus*) cultivars or breeding lines were tested for yield potential in nine different environments throughout the Pacific Northwest region of the United States. The analysis of variance was found to be that shown in Table 7.8.

From this we have detected highly significant differences between sites (locations) and cultivars; and a highly significant interaction between sites and cultivars.

However, when yield data from each location were analyzed separately we find that the error mean squares from each analysis of variance differ. This might have been expected given the large difference between the average yield at each location. The question is whether these differences are significant.

A significant χ^2 value ($p < 0.001$) is found from a *Bartlett analysis* indicating that these error variances are significantly different. In this case, therefore, we are violating one of the least robust assumptions on which the analysis of variance is based.

This is not uncommon and the question would then arise '*well, what can I do about a situation where there is heterogeneity of variances?*' The only practical solution to overcome this is to transform the data. The most common transformations are square root, log, and to transform the data to standardized normal distributions with mean zero and variance of one.

Given that a significant interaction is detected in the analysis of variance, and this is not due to a scalar effect, then it can often be useful to have additional

Table 7.6 Degrees of freedom and expected mean squares from an analysis of variance with g genotype entries, grown at l locations, and r replicates at each location. In this analysis it is assumed that locations are random effects.

Source	d.f.	EMS
Locations (l)	$l - 1$	$\sigma_e^2 + g\sigma_{rwl}^2 + rg\sigma_l^2$
Replicates w locations (r)	$l(r - 1)$	$\sigma_e^2 + g\sigma_{rwl}^2$
Genotypes (g)	$g - 1$	$\sigma_e^2 + r\sigma_{gl}^2 + rl\sigma_g^2$
Genotypes × locations	$(g - 1)(l - 1)$	$\sigma_e^2 + r\sigma_{gl}^2$
Error	$l(r - 1)(g - 1)$	σ_e^2

Table 7.7 Degrees of freedom and expected mean squares from an analysis of variance with g genotype entries, grown at l locations over y years, and with r replicates at each location. In this analysis it is assumed that locations and years are random effects.

Source	d.f.	EMS
Years (y)	$y - 1$	$\sigma_e^2 + gy\sigma_{rwly} + rgy\sigma_{lwy}^2 + rgl\sigma_y^2$
Locations w year (l)	$y(l - 1)$	$\sigma_e^2 + g\sigma_{rwly}^2 + rg\sigma_{lwy}^2$
Replicates w loc. and years (r)	$yl(r - 1)$	$\sigma_e^2 + g\sigma_{rwly}^2$
Genotypes (g)	$g - 1$	$\sigma_e^2 + r\sigma_{glwy} + rl\sigma_{gy}^2 + rly\sigma_g^2$
Genotypes × year	$(y - 1)(g - 1)$	$\sigma_e^2 + r\sigma_{glwy}^2 + rl\sigma_{gy}^2$
Genotypes × LwY	$y(g - 1)(l - 1)$	$\sigma_e^2 + r\sigma_{glwy}$
Error	$yl(r - 1)(g - 1)$	σ_e^2

Table 7.8 Degrees of freedom, sums of squares, mean squares and f-ratio values from an analysis of variance where 20 spring canola (*B. napus*) breeding lines were tested for yield potential at nine different environments throughout the Pacific Northwest region of the United States.

Source	df	SS	MS	F
Sites	8	220 698.7	27 581.3	133.6***
Reps w sites	27	5575.5	206.5	58.8***
Cultivars	19	2602.5	89.7	9.0**
S × C	151	1499.4	9.9	2.8***
Error	513	1801.7	3.5	

information regarding the partition of the interaction variance.

Interpreting G × E interactions

The idea of breaking the G × E interaction into several components is entirely missing from the simple analysis of variance table shown above. In the G × E context a method of partitioning this interaction was suggested by Yates and Cochran in 1938, although this was largely neglected for 20 years. In a paper these two statisticians stated 'the degree of association between varietal differences and general fertility (as indicated by the mean of all varieties) can be further investigated by calculating the regression of the yields of the separate varieties on the mean yields of all varieties'.

This is, ge_{ij} in the model equation (7.1) above is regressed onto e_j. Therefore:

$$ge_{ij} = \beta_i e_j + \alpha_{ij} \qquad (7.2)$$

where β_i is a linear regression coefficient for the ith genotype and α_{ij} is a deviation from regression. Using a combination of equations (1) and (2) we can write:

$$y_{ijk} = \mu + g_i + (1 + \beta_i)e_j + \alpha_{ij} + E_{ijk}$$

An analysis of variance can be used to partition the genotype × environment interaction into *heterogeneity of regression* (i.e. that the regression slopes of the different genotypes have different slopes) and *deviation from regression* (i.e. that the relationship between genotypes and environments is not explicable by linear regression). In an analysis of variance with g genotypes and l environments the partition of interaction sum of squares would give $(g - 1)$ degrees of freedom for heterogeneity of regression and $(g - 1)(l - 2)$ degrees of freedom to deviations from regression.

In the analysis, each of these terms can be compared in an F-test on division by the error mean square. The heterogeneity of regression can be further compared with the deviations from regression to see if it accounts for a significant part of the observed interaction.

This particular approach is called *joint regression analysis*. Despite some major theoretical difficulties with the analytical technique, it has been widely used by plant breeders to determine the stability of genotypes in the advanced stages of selection.

From the analysis each genotype has an average performance (mean over all environments) and a regression coefficient $(1 + \beta_i)$. The regression coefficient can be used to determine the stability of different genotypes over environments. Genotypes which have high $1 + \beta$ values are said to be more responsive to environment

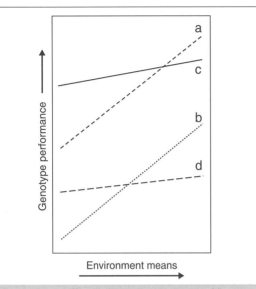

Figure 7.11   Different phenotypic response (yield) to environment as measured by average performance and regression coefficient of performance against average environment performance where (a) has high average yield and high $1 + \beta$ response, (b) low average yield and high $1 + \beta$, (c) high average yield and low $1 + \beta$, and (d) low average yield and low $1 + \beta$.

than those with low $1 + \beta$ values. From the analysis there can be four types of genotype:

- High average performance and high $1 + \beta$ value
- Low average performance and high $1 + \beta$ value
- High average performance and low $1 + \beta$ value
- Low average performance and low $1 + \beta$ value

The four types of genotype are shown diagrammatically in Figure 7.11. Obviously all plant breeders would like to develop cultivars which have high average performance and low $1 + \beta$ values as these genotypes would produce high yield in all environments. Unfortunately performance and regression coefficient are usually related with high expression associated with high sensitivity.

Major criticism of joint regression analysis is that the regression is carried out between the performance of each genotype at each site onto the average performance of all genotypes at that site. Obviously these two are not independent as the genotype performance is *a factor in determining the site mean*. To avoid this it has been

suggested that the genotype being regressed should be omitted from calculating the site mean, although this has been shown to make little difference to the slopes obtained. Another way to avoid this is to carry out regression of the breeding lines onto the performance of one (or more) control cultivar (say one which is known to have stable performance over locations and years).

Other methods are available to examine $G \times E$ interactions in breeders' trials. Genotypic stability over environments can be examined by comparing the variance of each genotype over the different environments. This method can produce very similar results to joint regression analysis with increasing variation over environments related to less stability.

Genotype means and variances over environments can also be used simultaneously to predict the frequency that test entries would exceed specified target values (often the value of controls in the trials). One advantage of the probability prediction method is that it provides a single datum for each test entry to be evaluated (i.e. the probability that a genotype will produce a yield greater than 2000 kg/ha). A second, perhaps more important feature, is that this method can be expanded to cover several traits simultaneously. This is achieved by estimating the average performance for each trait, the variation over environments for each trait and the covariance between traits and environments. Probabilities are calculated by evaluating the area under a univariate or multi-variate normal frequency distribution (see cross prediction later).

Finally, another form of analysis called **Additive Main Effects and Multiplicative Interaction** (AMMI) has been suggested as suitable by several researchers. An AMMI analysis partitions the residual interaction effects between genotypes and environments by principal component analysis. The technique therefore involves a multi-variate transformation of the residuals (after the main effects of genotypes and environments have been removed) within a two-dimensional table of genotypes and environments. Provided the first and second eigen vectors (transformations) account for a large proportion of the residual data, two-dimensional graphic inspection of eigen values from the first and second eigen vectors, for different locations and genotypes can be used to examine the interaction effect. The theory of AMMI is outside the scope of this book although readers should be aware that this option of interpreting data is available.

Selection

Understanding data and interpreting results from multi-location trials are a great aid in selection from such trials. Factors of importance in selection will include genotypic stability (or lack of) and correlation between characters over environments and between traits within a single location.

Finally, efficient selection with the very large data sets that are common in the advanced selection stages can be helped by inspection of performance ranking. Indeed it has been shown that summing the rank of individual traits (or single traits over environments) will produce similar results to multi-variate probabilities. Summing ranks can be considerably easier to compute than finding the area under an *n*-dimensional normal frequency distribution.

CROSS PREDICTION

The number of breeding lines discarded in a plant breeding programme is inversely proportional to the number of selection rounds. In the early stages there are many thousands of lines evaluated with a low proportion that are selected for testing at the intermediate stage. At the advanced stage a few surviving lines are tested with great intensity with only a few lines being discarded after each selection stage.

A number of researchers have shown that selection in the initial stages (that period where the greatest proportion of genotypes, and hence the greatest genotypic variation are discarded) is the most ineffective stage at identifying the most desirable lines. At the early generation stage, selection has been shown to result in, at best, only a random reduction in the number of genotypes within the breeding scheme. Some advances (particularly for qualitatively inherited traits) have shown a response to initial selection although some have shown a negative response where the best phenotypes under conditions of early generation selection have been shown to be those least likely to become commercial cultivars.

Therefore, the early generation selection stage of a plant breeding programme is often very ineffective in terms of selection. This inefficiency is in part simply due to low heritability in performance in the early selection stages compared with those in more advanced levels.

Low inefficiency is, in part, the result of:

- The inaccuracy of selecting on small plots (often single plants) because of the error variance and sampling variation along with competition effects of surrounding plots. This is not helped by the inability to adequately replicate and/or randomize the vast number of genotypes involved at the earliest selection stages
- Selection under atypical conditions which do not mimic plant spacing etc. that would be common in commercial production
- Selection amongst highly heterozygous lines where genotypic worth can be severely masked by dominance effects (inbreeding species only)

Selection in the early stages is most ineffective for quantitatively inherited traits such as yield, quality and durable disease resistance. If early generation selection is purely a random (or near-random) reduction in the number of lines which are to be tested at the intermediate stage then it is questionable whether this operation will merit the time and resource to complete the task. It has therefore been suggested that a more effective protocol would result from growing fewer breeding lines and doing no selection at the earliest stages. The reduced effort and resource at the early generation stage could therefore be used more efficiently to screen more genotypes at the intermediate stage (where efficiency due to replication and larger plots is more effective).

An alternative method of reducing the numbers involved in early to intermediate selection is available. This procedure involves the identification of the most attractive cross combinations from the many that would be possible, assuming that there is greater probability of obtaining a successful cultivar from the most desirable cross combinations. Having identified the '*best*' crosses then maximum effort and resource can be directed to screening individual recombinants from within these lines, while the '*poorer*' crosses are completely discarded. This process is called *cross prediction*.

Univariate cross prediction

As we noted briefly earlier, methods of predicting the properties and distribution of recombinant inbred lines (derived by inter-mating homozygous parents) using early generations of crosses have been proposed by Jinks and Pooni.

They have shown that for any continuously varying character, the expected mean and variance of all possible inbred lines, derived by inbreeding following an initial cross between two homozygous parents, can be specified in terms of the components of means and variances as specified by biometrical genetics. For example, if an additive-dominant genetic model of inheritance proves adequate, the expected mean is m, the mid-parent value, and the expected variance of the inbred sample is V_A.

From the predicted mean and variance we can determine many of the properties of the recombinant inbred lines that can be derived in a pure-line breeding programme based on the performance of generations in the early generation stages. In addition, the relative probabilities with which different pair-wise crosses will produce inbred lines with particular properties can also be predicted and hence used as a selection criteria for reducing the number of breeding lines in a plant breeding programme.

The crosses which show highest probability of producing desirable recombinants can therefore be identified from those with a lesser chance of producing desirable lines. Rather than selecting individual genotypes at the early generation stage, the number of surviving lines can be reduced by selection of the superior cross combinations. Similarly, if the probability of a desirable recombinant is known from a particular cross, then this value can be used to determine the number of recombinants which need to be evaluated to ensure that 'at least one' is found. When a single trait is examined, this procedure of estimating and using genetic parameters is called ***univariate cross prediction***.

Univariate cross prediction has been applied to a number of inbreeding species based on the initial work of Jinks and Pooni with *Nicotiana rustica*. Predictions of the proportion of recombinant inbred lines that will transgress a predefined target value (T) are based on the evaluation of the integral:

$$\int_T^\infty f(x_i)\,\mathrm{d}x_i$$

where the variate of interest is normally distributed and the function $f(x_i)$ is based on m, the mean of all possible inbreds for a character and A, the additive genetic variance for the character (Figure 7.12).

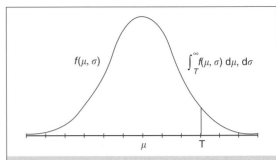

Figure 7.12   Illustration of Univariate cross prediction technique.

Estimation of m and V_A

The additive genetic components of the expected variance (V_A) can be estimated from a number of different sources initiated by a cross between two pure-breeding lines. Methods that have proven reliable include:

- Producing a sample of inbred lines from a pair-wise cross between two parents by rapid cycle single seed descent or doubled haploid techniques
- The standard P_1, P_2, F_1, F_2, B_1 and B_2 families
- A triple test cross, where a random sample of the F_2 from each parent cross-combination is backcrossed to P_1, P_2 and the heterozygous F_1 (The theory of the triple test cross is beyond the scope of this book.)
- Evaluation of a random sample of F_3 families

If a sample of inbred lines from a number of different crosses are grown in properly designed assessment trials, it is possible to estimate m, the average performance of all possible inbred lines from each cross and A, the additive genetic variation for each cross. The average performances of the inbreds are a direct estimate of m, and the variance between inbreds (σ_g) in the sample is a direct estimate of A after error variance has been removed.

It is possible to calculate the proportion of lines expected to transgress a predefined target value (T) by using:

$$\frac{T - m}{\sigma_g} \quad \text{or} \quad \frac{m - T}{\sigma_g}$$

depending on whether the predictions are for values greater than (or equal to), or less than (or equal to) the target value set. Where T is the target value, m is the mean of all inbred lines and σ_g is the genetic standard deviation.

Following the calculation of the probability integral from the predicted equations, the expected proportions of transgressive segregants can be obtained from tables of the normal probability integral.

In cases where it is easy to obtain a sample of inbred (or near-homozygous) lines from a large number of crosses (i.e. by using a doubled haploid techniques or rapid cycle single seed descent) then this method will produce excellent predictions as there are no dominance effects to complicate estimation.

In many instances, however, it is not possible to produce inbred lines quickly and cheaply on a practical level in a breeding programme, and so cross prediction will involve estimating genetic means and variances from early generations of crossing designs.

Although the triple test cross will provide breeders with the best estimate of additive genetic means and variances it requires a great deal of time and effort to complete. A similar effort will be required to obtain these estimates using the standard P_1, P_2, F_1, F_2, B_1 and B_2 method. Both these mating designs therefore have merit for genetic investigation but may have limited use in a practical plant breeding situation where many hundred of cross combinations need to be screened.

Evaluation of a random sample of F_3 families from each cross under investigation offers a more practical approach. Approximate genetic parameters can be estimated from the mean of a random sample of F_3 families and from variation between families derived from a common cross.

For example, consider a single cross ($P_1 \times P_2$). Then F_1 seed would be grown to produce F_2 single plants. A random sample of these would be harvested and a single plot grown from each of (say 20 to 25) single plant plots. These plots would be evaluated to obtain the average performance of the families and variation between families would also be estimated.

The mean (average performance of families) of the F_3 plots would be:

$$m + 1/4d$$

and the true variance of the F_3 family means (σ_{F3}^2) would be:

$$1/2V_A + 1/16V_D$$

assuming that the additive-dominance model of inheritance is adequate to describe the character of interest. Therefore the following approximations can be made:

mean of F_3 families = mean of all possible inbred lines

$$2\sigma_{F_3}^2 = V_A$$

Both estimates will of course be accurate only if dominance effects are relatively small in comparison to additive effects. In cases where $[d]$ is large, then the average of all possible inbred lines can be estimated by growing the parental lines in the prediction trial and estimating m as $(P_1 + P_2)/2$. Alternatively, when $[d]$ and D are large then they can be estimated by including a bulk sample of the F_1 or F_2 in the prediction trial. This latter option will of course also offer a better estimate of m.

If the parental lines are included in the cross prediction trial in which the F_3 families are evaluated, then it is also possible to carry out a crude scaling test to determine the dominance components and effect from:

$$(P_1 - P_2)/2 - F_3$$

Using these predictions of the additive genetic mean and the additive genetic variance we can predict the probability that a single inbred line selected at random will exceed a predefined target value (T) by:

$$\frac{T - \text{mean } F_3}{\sqrt{(2\sigma_{F_3}^2)}}$$

Similarly, the probability that a single inbred line, taken at random, would be less than a predefined target value (T) would be:

$$\frac{\text{mean } F_3 - T}{\sqrt{(2\sigma_{F_3}^2)}}$$

Following the calculation of the probability integral from the predicted equations, the expected proportions of transgressive segregants can, as above, be obtained from tables of the normal probability integral.

Setting target values

A number of different options are available in setting target values on which the predictions are based. These include:

- To include a set of control cultivars in the same experiments where the genetic parameters are estimated and using the means (and variances) to set target values which would be needed by a new cultivar

- To include all parents used in crosses and setting target values of either P_1 (the parent with highest expression) or $(P_1 + P_2)/2$, the average parent value or m
- If the F_1 progeny are included in the evaluation trial then the heterotic F_1 values are often used as target values in cross prediction

Predicted from number in sample

When a sample of inbred lines is produced then a second option of determining the frequency of particular recombinants has also shown promise. This method simply involves counting the number of inbreds within the sample which exceed the given target value.

Obviously, the accuracy of this method will be directly proportional to the sample size of inbreds used on which the counts are made. Several researchers have shown, however, that even relatively small samples (around 25 lines) can still provide useful prediction results.

Use of cross prediction in clonal crops

Initially cross prediction was not considered as a tool in the breeding of clonal crops. This may perhaps have been due to the fact that heterozygosity is not a problem in the selection procedure. In other words, although the initial seedlings in say, a potato breeding programme, are all genetically unique and highly heterozygous, they are subsequently multiplied clonally and so are fixed in the sense that they are the genotypes which can be commercially exploited. However, it has been shown in a number of clonal crops that early generation selection suffers from all the inefficiencies found in inbred cultivar development.

In clonal crops, there are, just as many difficulties incurred in trying to identify desirable lines in the early generation stages (i.e. seedling stage and first clonal year stage) where only single plants are evaluated.

At the Scottish Crop Research Institute research has shown severe inefficiencies in the method of selection used in the early generations. Work resulting from this prompted an examination of cross prediction methods which might prove an effective alternative to the recurrent phenotypic selection used. A sample of 25 seedlings from the 200 grown from each of 8 cross

Table 7.9 Progeny means of Breeders' Preference ratings and within progeny variances (σ_p), estimated on 25 progeny from each of eight potato crosses (C1 to C8), and the univariate probability that a genotypes chosen at random from each cross will exceed a Breeder's Preference rating greater than 5, on the 1 to 9 scale.

Cross	Mean	σ_p	Predicted >5
C1	4.36	1.52	0.337
C2	4.01	1.65	0.274
C3	3.61	1.50	0.176
C4	4.17	1.23	0.251
C5	3.04	0.91	0.015
C6	3.68	1.52	0.192
C7	4.21	1.36	0.281

families (C1 $\cdots$ C8) were evaluated for breeders' preference (a visual assessment of commercial worth on a 1 to 9 scale with increasing value attributed to increasing commercial worth) by four breeders independently. From the average assessment of each genotype, Table 7.9 shows the progeny means and within progeny variances (σ_p) used to determine the frequency of clones which would exceed a preference score of greater than five on the one to nine scale.

Seed tubers from a sample of 200 clones from each cross were increased without selection to a stage where a large amount of field grown seed tubers were available and the 1600 genotypes evaluated for three years in a breeding programme. Selection of the populations was based on all the characters (yield, quality and appearance) which are normally assessed in the breeding scheme. Table 7.10 shows the number of selected clones which survived four, five and six rounds of selection from each cross. Also shown is the rank of each cross based on the 25 seedlings and cross prediction of breeders' preference.

Obviously, there were more highly desirable clones from cross C1, C2 and C7 which were ranked first, second and third on the univariate cross prediction of glasshouse grown seedlings.

In the potato, cross prediction was investigated with higher numbers of crosses (204) in a similar way as a result of this first study. Results from the larger study were in agreement with those shown above.

Table 7.10 Relative ranking of eight potato crosses (C1 to C8) based on cross prediction of Breeder's Preference that genotypes chosen at random from each cross will exceed a Breeder's Preference rating greater than 5, on the 1 to 9 scale and the number of selected breeding lines from each cross that was selected in the 4th, 5th and 6th selection stage in the breeding program at the Scottish Crop Research Institute.

Cross	Rank	Selected to stage		
		Four	Five	Six
C1	1	15	3	2
C2	3	9	3	2
C3	6	1	0	0
C4	4	2	0	0
C5	8	1	0	0
C6	5	11	6	1
C7	2	12	7	3
C8	7	0	0	0

This has prompted several other breeding organizations to change the means by which they reduce clonal numbers in the early generations of potato and sugarcane breeding schemes.

Use of normal distribution function tables

The area under a unit normal distribution (a normal distribution with mean of zero and variance of one) is frequently tabulated in statistical tables. It is common for the area to be given from $-\infty$ to the required target value (T). The whole area from $-\infty$ to $+\infty$ is of course equal to 1, so the area from T to $+\infty$ can be obtained by subtracting the table value from 1.

For example, from a given cross (A × B) estimates were obtained of the mean ($m = 12.0$), genetic variance ($\sigma_g = 16.0$, therefore $\sigma_g = \sqrt{\sigma_g^2} = \sqrt{16} = 4$). What would be the probability that a recombinant will exceed a set target value of 14? To solve this we have:

$$\frac{T - m}{\sigma_g} = \frac{14 - 12}{4} = 0.5$$

Using the unit normal distribution tables and a value of 0.50 we have the probability of $-\infty$ to T to be equal to 0.6915 (from table of unit normal distribution). The actual probability we want is $1 - 0.6915 = 0.3085$.

Therefore given the above genetic parameters we would expect that 30.85% of all possible recombinants from the cross will have a greater (or equal) value than the target value.

Consider the same set of parameters ($m = 12$, $\sigma_g = 4$) but now with a target value of 11 (i.e. a target value less than the progeny mean). We now have:

$$\frac{T - m}{\sigma_g} = \frac{11 - 12}{4} = -0.25$$

Looking this value up from the tables we have a probability value of 0.5987. In the example above we then subtracted this value from one to obtain the correct probability. In this case however, this $(T - m)/\sigma_g$ has a negative value as so our required probability is $1 - (1 -$ table value) which is in fact, simply the value obtained from tables.

In summary, four possibilities exist:

- If the probability that a recombinant is to **exceed** a target value and $(T - m)/\sigma_g$ is **positive** then the probability is $1-$ the tabulated value
- If the probability that a recombinant is to **exceed** a target value and $(T - m)/\sigma_g$ is **negative** then the probability is simply the tabulated value
- If the probability that a recombinant is **less than** the target value and $(m - T)/\sigma_g$ is **positive** then the probability is simply the tabulated value
- If the probability that a recombinant is **less than** the target value and $(m - T)/\sigma_g$ is **negative**, then the probability is $1-$ tabulated value

Univariate cross prediction example

To consider further possible problems involved in selection of the 'best' cross combinations consider the following example.

Below are shown the means and genetic variances of crop yield (t/ha) of four barley crosses (A, B, C and D). Also grown in this prediction trial were five control lines. These controls are all commercially grown cultivars predominating over the region where new varieties are to be grown. The average yield of the controls was 21 t/ha and the variance of the controls was 7.5 t/ha. Which of the crosses should have greatest emphasis in a breeding programme where high yield is the major selection criteria?

Cross	mean	Genetic variance
A	20.0	24.135
B	22.0	8.111
C	21.5	19.245
D	18.0	26.051

First it should be decided if selection is based only on the mean performance of the crosses. If this is the case then the answer is quite simple. Greatest emphasis should be made with cross B, followed by cross C. The remaining two crosses perhaps should be discarded as their average performance is less than the average of the control cultivars.

It should be noted that the four crosses have different mean yield values but also there are large differences in the genetic variance (σ_g^2). Would our decision now change if we consider the '*best*' cross based on the mean and variance?

First it is necessary to set a target value on which the prediction is to be based. As there were a number of commercial cultivars included in the cross prediction trial it may be useful to use as the target value the average performance of the controls (21 t/ha).

When this target value is used the four crosses were estimated to have A = 42.4%, B = 63.7%, C = 54.4% and D = 27.8% of their progeny to be greater (or equal) to the mean of the control entries. Again if these were the criteria used then greatest emphasis should be put on cross B, followed by cross C, A and lastly D (the same order as when only the means were used).

If this were an actual breeding scheme, however, it may be several years before a selected genotype from any of these crosses becomes a commercial cultivar in agriculture. It may therefore be wise to set our target higher than the controls as it might be expected that in several years time then newer and higher yielding lines will be available. As the variance of the controls is available we can use this to set a target value which is the mean of the controls plus the standard error of the controls (i.e. 21.0 + 2.74) which would be approximately 24 t/ha.

With this target value we have A = 20.90%, B = 24.20%, C = 28.42% and D = 11.90%. Now there has been a change in the cross rankings (in parenthesis, Table 7.11) with cross C now giving the highest probability of a lines exceeding 24 t/ha. Cross B is now ranked second but there is little difference between the probabilities of cross B and cross A.

If the target value is further increased (say to the control mean plus twice the control standard error, we would have a target value approximately equal to 26. With this target value the ranking of the four crosses is C, A, B and D. Now cross A has a higher probability of a genotype exceeding the target value (11.12%) than cross B (8.08%).

In conclusion therefore, univariate cross prediction is based on the mean and genetic variance of a cross. When target values are relatively close to the progeny mean values then not surprisingly the mean of each cross will be a large factor in the cross prediction. As target values are increased then the genetic variance becomes a more important factor in determining the probability of desirable recombinants. Finally, in the above example it should be noted that it is the genetic standard error (σ_g) that is used in the estimate and not the genetic variance (σ_g^2). Even when there are large differences

Table 7.11 Progeny mean, genetic variance and genetic standard deviation of progeny from four different parent cross combinations, and the probability that a genotypes chosen at random from each cross will exceed a specific target yield.

Cross	Mean	σ_g^2	σ_g	$T = 21$	$T = 24$	$T = 26$
A	20.3 (3)	24.13	4.91	0.424 (3)	0.209 (3)	0.111 (2)
B	22.0 (1)	8.11	2.85	0.637 (1)	0.242 (2)	0.081 (3)
C	21.5 (2)	19.24	4.38	0.544 (2)	0.284 (1)	0.153 (1)
D	18.0 (4)	20.05	5.10	0.278(4)	0.119 (4)	0.058 (4)

in genetic variance (compare cross B and cross D) the cross with highest mean value was always the better choice for further breeding work, despite the high variance of cross D. However, if the target was taken to a greater extreme, then the relationship would cease to hold true, and extremely *'good'* genotypes is what breeders are usually trying to identify.

Multi-variate cross prediction

Despite the usefulness of univariate cross prediction in determining the frequency of desirable recombinants that would transgress a given target value, its use is limited because only a single character can be evaluated. As we noted many times already, usually a new cultivar will not be successful because of high expression in a single character, but rather it needs to express an overall improvement in a number of morphological, pathological and quality characters combined with high productivity.

The problem of selecting the most desirable cross combinations can partially be overcome by considering a variate such as breeders' preference, which is based on a visual assessment of several characters simultaneously by a breeder. Indeed breeders' preference scores have been shown to give very similar results to multi-variate index selection schemes.

Visual inspection of several characters simultaneously, to result in a single overall rating for each individual has several limitations. In potatoes this form of assessment has been shown to have advantageous features when used in a plant breeding selection scheme. Breeders' preference scores in potato breeding are highly related to actual yield, number tubers per plant, tuber size, tuber conformity, tuber disease and absence from defects. It has been shown that this type of evaluation does not have such a good agreement with other important characters such as seed size, disease resistance, yield etc. Similarly it is not possible to combine characters which are expressed at different times. For example it is difficult to consider pre-harvest characters such as flowering time plant height or maturity if preference scores are recorded at harvest. In addition it is difficult to combine morphological characters such as yield along with quality characters that may be assessed in a laboratory at a later stage. Thus it may be necessary to consider selection for more than a single trait.

If more than one trait is to be considered in cross prediction studies it is possible to treat each independently, carry out univariate cross prediction on each character and examine the probabilities obtained to make decisions on the 'best' crosses. This would of course ignore the fact that the different traits are inter-related (correlated) and that the relationship between the traits is constant over all crosses involved. This may cause problems and so it may be necessary to expand the univariate procedure to cover several different traits simultaneously.

Univariate cross prediction is based on evaluation of the normal distribution function determined by the mean and genetic variance of each cross and a chosen target value (T), i.e.:

$$\int_T^\infty f(x_i)\mathrm{d}x_i$$

Suppose that two characters are to be considered. The bi-variate normal distribution of the data from these two traits can be described by the mean of each character (m_1 and m_2), the genetic variance of each character (σ_1^2 and σ_2^2) and the correlation between the characters (τ). Given these five parameters it is possible to estimate the proportion of recombinants from the cross that will transgress a given target value for character 1 (T_1) and simultaneously transgress a second target value (T_2) for the other trait. This probability is given by:

$$\int_{T_1}^\infty \int_{T_2}^\infty f(x_1, x_2)\mathrm{d}x_1, \mathrm{d}x_2$$

where the function $f(x_1, x_2)$ is a bi-variate normal distribution function based on the mean of both traits, the variance of both traits and the correlation between traits.

It is easy to extend this to cover n different traits by evaluation of the integral:

$$\int_{T_1}^\infty \int_{T_2}^\infty \cdots \int_{T_n}^\infty f(x_1, x_2, \ldots, x_n)\mathrm{d}x_1, \mathrm{d}x_2, \ldots, \mathrm{d}x_n$$

In this case the function $f(x_1, x_2, \ldots, x_n)$ is a multi-normal distribution function based on the mean ($m_{1,n}$) of all n traits, the genetic variance ($\sigma_{1,n}^2$) of all n traits and the genetic correlation ($\tau_{i,n}$) between all n traits.

Given the various means, variances and correlations it is possible to obtain bi-variate and tri-variate probability estimates from statistical tables. These tables are, however, not commonly presented in standard statistical tables (as for example the ones that usually show unit normal distribution function, *t*-tables, χ^2-tables or F-tables). In addition use of the tables that do exist can be complex and would require detailed description.

Parameters used in multi-variate prediction are estimated using the same design types (i.e. triple test cross, F_3 prediction) that were explained previously for univariate predictions.

When it is necessary to estimate multi-variate probabilities, computers offer an easier alternative. Computer software is available (although not commonly) which projects a probable value, when the means, variances, correlations and target values are entered. To our knowledge there is software which can handle upto seven traits simultaneously – how the software manages this need not detain us here!

Similarly, it is beyond the scope of this book to try to explain in more detail the theory of estimating these probabilities. It is sufficient to understand the basic concept and to be aware of the usefulness of the procedure as applied to cross prediction techniques. You should, however, be aware that the procedure exists and that multi-variate cross prediction can offer a powerful tool to selection in plant breeding.

Example of multi-variate cross prediction

The eight crosses (C1 ⋯ C8) that were evaluated for breeders' preference (see earlier in univariate prediction) also had tuber yield, tuber size and number of tubers recorded for the 25 progeny from each cross. Tuber shape was also visually assessed. The means and variances of each variate were estimated along with the correlation between traits for each cross. Based on these statistics the probability that genotypes would exceed target values for each character simultaneously was estimated using a computer software package called POTSTAT. The relative ranking of the multivarate predicted values (MV.rank) are shown in Table 7.12 along with the ranking of the univariate cross prediction of breeders preference (UV.rank) and the frequency of desirable clones selected from a large sample in the fourth, fifth and sixth round of selection.

There is good agreement between the multi-variate predictions, based on four traits and the univariate prediction based on breeders' preference. Therefore the

Table 7.12   Relative ranking of eight potato crosses (C1 to C8) based on multivariate cross prediction (MV-rank) and univariate cross prediction (UV-rank) of Breeder's Preference that a genotypes chosen at random from each cross will exceed a Breeder's Preference rating greater than 5, on the 1 to 9 scale and the number of selected breeding lines from each cross that was selected in the 4th, 5th and 6th selection stage in the breeding program at the Scottish Crop Research Institute.

Cross	MV.rank	UV.rank	Selected to stage		
			Four	Five	Six
C1	2	1	15	3	2
C2	2	3	9	3	2
C3	6	6	1	0	0
C4	4	4	2	0	0
C5	8	8	1	0	0
C6	5	5	11	6	1
C7	1	2	12	7	3
C8	7	7	0	0	0

preference scores were highly related to yield, number of tubers, tuber size and tuber shape. There was also very good agreement with the predicted worth of each cross and the number of clones which indeed show commercial value in the advanced selection stages.

Observed number in a sample from each cross

It is possible to obtain good multi-variate probability estimates by observing the frequency of individuals in a small sample that exceed given target values. The difficulty in using observed frequencies is related to sample size. The accuracy of the predictions will be directly related to the sample size examined. When the frequency of desirable recombinants is low (i.e. when large target values are used) then larger samples will need to be examined.

Similarly, if there are low correlations between traits of interest sample sizes will need to be relatively large to predict effectively.

Use of rankings

It has been noted, above, in the univariate case that the relative importance of the different parameters can affect the results of prediction. In multi-variate prediction three types of parameter are used, means, variances and correlations. In cases where the progeny means predominate in the prediction equations, which is often the case, then very good estimation of progeny worth can be obtained by summing the relative rankings (based on the phenotypic mean of the cross) for each of several traits.

Consider the potato example shown above where four traits, yield, tuber size, tuber number and shape were used to assess progeny worth of eight potato crosses. The ranking of the eight crosses based on the multivariate normal probability (MVP) and those obtained by summing the relative ranking of each character were:

Cross	Sum Rank	MVP
C1	1	2=
C2	2=	2=
C3	5	6
C4	4	4
C5	8	8
C6	6	5
C7	2=	1
C8	7	7

As can be clearly seen, ranking each individual trait and then summing the rankings of each cross can be a good estimate of the commercial worth of different cross combinations.

PARENTAL SELECTION

Selection in a plant breeding programme takes two forms:

- Selection of superior parents
- Selection of desirable recombinants which have resulted from inter-mating chosen parents

Selection of the desirable recombinants has been covered in the foregoing sections of this book. We will now consider parental selection.

Parents used in plant breeding programmes are chosen from a wide source of possible genetic material. In general, however, parents are of three different types:

- Unadapted or relatively unadapted genotypes which possess one (or more) character which is not available

within more cultivated types (i.e. parents from plant introduction accessions and germplasm gene-banks)
- Adapted genotypes which may be new (or old) cultivars from other breeding programmes
- Genotypes which have been selected from within the breeding programme. Often these lines will become new cultivars but it is not uncommon that advanced selections (which have only a few slight defects that would render cultivar introduction infeasible) are used as parental lines

It may appear strange that recombinant selection was discussed prior to parental selection (i.e. putting the cart before the horse). In actual practice there is no definite order of either selecting parents or selecting offspring. A large majority of parents used in plant breeding scheme are derived from selections within the breeding programme.

Parental selection is therefore a cyclic operation where parents are selected, inter-mated, recombinants screened from segregating populations and these, in turn, are used as parents in the next round of the scheme.

In deciding which parents are to be used in a breeding scheme there are two types of evaluations possible:

- Phenotypic evaluation
- Genotypic evaluation

This information could have been derived from experiments or assessment trials carried out within the breeding scheme or by other organizations (i.e. germplasm databases).

Phenotypic evaluation

Phenotypic evaluation is often the first stage of parental selection. New genetic material is continually being added to the available parental lines within a plant breeding programme.

It can be of great benefit to a breeder, and will add increasing knowledge of possible new parental lines, to grow parental evaluation trials. When a potential new parent is made known often the information of commercial worth is lacking. Information may be available from a database management scheme although often these data are related to performance in different regions to the target region of the breeding programme.

Phenotypic parent evaluation trials can be carried out at relatively low cost. When many new parents are to be assessed then specific trials can be arranged. These trials should be organized with the same criteria of good experimental design that other evaluation requires. In cases where only one or two new parents are to be considered it is often useful to include these genotypes as controls in one of the breeding trials.

Genotypic evaluation

However, although often only after a new parent has proven to have some merit on its phenotypic performance is a more detailed examination of genotypic worth carried out, the possibility that a valuable genotype (in terms of becoming a parent) might hide within a poor phenotype still exists. Nevertheless, because of limited resources and a lesser probability of a poor phenotype proving to be a good genotype, most effort is devoted to further evaluating proven material to determine the true value of the parent in cross combination.

The most common means to determine the genetic potential of new parental lines is to examine a series of progeny in which the new parent features as one of the parents. From these studies it is possible to determine the *general combining ability* of a series of different genotypes and to use this information to select the most desirable parental lines.

General combining ability is an indication of how the progeny from a particular genotype crossed to a range of other genotypes responds. The most effective means of determining general combining ability is by diallel crossing designs, where the variation observed in the diallel table is divided into general combining ability of the parents used and specific combining ability (all variation which cannot be explained by an additive model of parental values). But as noted before, this does limit the number of lines that can be examined.

General combing ability can be estimated from other crossing designs. The simplest of these involves evaluating the progeny that are produced by crossing the potential parent with one or more tester lines. Tester lines are chosen because of past experience in producing worthwhile results. For example, a new parent may be crossed to a genetically productive genotype and also to one with little genetic worth. The contribution of the parent can be observed by examination of the offspring from the crosses.

General combining ability can also be estimated using North Carolina crossing designs. These, as noted earlier, are of two forms:

- *North Carolina I designs*, where a number of parent genotypes are crossed to one or more tester lines. In these designs it is not necessary to hybridize each parent to a common set of tester lines. From the design an analysis of variance can estimate general combing ability of parents, which is tested for significance against the testers within parents' mean square.
- *North Carolina II designs*, where a number of parent genotypes are crossed to one or more tester and each parent is crossed to the same tester lines. In this case an analysis of variance can partition the total variation into differences between specific combining ability of the parents, combing ability of the testers and an interaction term (parents × testers) which indicates specific combing ability.

In addition to the statistical analysis of diallels and other crossing designs more information can be obtained from genetic analysis. The most common means to achieve this is from a Hayman and Jinks' Analysis where within array variances (V_r) and between array covariances (W_r) are used to estimate the proportion of dominant to recessive alleles for a given character. Hayman and Jinks' Analysis can be used therefore to choose parents which have high phenotypic performance and with a high degree of dominant alleles.

When a suitable cross prediction scheme is employed in the early generation of a plant breeding scheme it is possible to use the cross prediction data to indicate which specific parents have the highest probability of producing desirable recombinants. A potential new parent is hybridized to a number of different genotypes and the progeny are examined to estimate the mean of all crosses in which the parent is used and the genetic variance of all crosses in which the parent appears. These data can be used in the same way as illustrated earlier in cross prediction.

Similar probabilities based on several traits simultaneously can provide useful indicators of the exact worth of a new parent without waiting several years to determine this potential from survivors in a selection scheme.

At the Scottish Crop Research Institute cross prediction at the seedling stage of the potato breeding programme became a standard practice. Each year between 200 and 300 crosses were evaluated in cross

Table 7.13 Univariate probability that a genotype taken at random from a segregating progeny with a common parent will have a Breeders' Preference greater than 4, on a 1 to 9 scale, relative raking of that probability, and proportion and ranking of genotypes that survive the 4th selection stage at the Scottish Crop Research Institute.

Clone	Cross prediction of preference > 4	Rank	Percentage of year 4 clones selected in year 7	Rank
Maris Peer	69.17	1	17.69	1
3683.A.2	62.57	2	11.76	2
Pentland Ivory	60.40	3	7.11	4
G.6755.1	59.74	4	6.29	5
Cara	57.34	5	10.95	3
8204.A.4	54.42	6	5.13	6
Pentland Squire	49.25	7	3.18	7
Dr Macintosh	47.37	8	0.00	8=
Self crosses	37.99	9	0.00	8=

prediction trials. The numbers of genotypes from those hybrid combinations with the highest probability of producing a new cultivar were increased while the less desirable cross combinations were discarded.

This scheme, in addition to providing information of the commercial potential of each cross combination, also was used to determine the suitability of individual parents. The progeny mean and genetic variance of each parent was used in the prediction estimation. In Table 7.13 are shown rankings of 9 parents based on this system along with the number of desirable recombinant lines which resulted from crosses involving the parents. Despite one or two changes in rank order there was good agreement in the predicted and observed indicators. The differences which were observed could be explained by morphological characters (i.e. Cara has a pink eye and there was positive emphasis to select these types) or pest preferences (i.e. Maris Piper has nematode resistance and only clones which possessed the resistance were continued irrespective of other characters).

Parental combinations

Having decided on a set of parental lines the next decision to be made is *how many crosses should be made* and *which combinations will yield best results?*

If there is a means by which large numbers of crosses can be evaluated then many crosses will yield better results than if only a few are tried. However, it should be noted that there is little to be gained by making more crosses than can be screened in an effective manner.

In a straightforward commercial context, and for a short-term objective, only a limited number of crosses are to be considered, one simple and effective strategy is to cross the best with the best. Therefore identify the phenotypically and genetically best parents, intercross these and select amongst their progeny.

Many breeders use the strategy of combining complementary parents. For example, to inter-mate a high yielding poor quality line with a low yielding but high quality line. In theory this type of combination could allow the selection of a high yield high quality recombinant. However, what is often achieved is an average yield with average quality. It is usually necessary to use some form of pre-breeding where the high yielding line is first crossed (or backcrossed) to a high quality line, parents are selected several times, and these are in turn used in final cross combinations.

Similarly when a character is introduced from a wild or unadapted genotype it may take many rounds of backcrossing to get the desired character into a commercial background before the trait is introduced into a new cultivar. This is of course, where some of the newer techniques of genetic transformation and marker assisted selection offer other alternatives to the breeder.

Germplasm collections

Germplasm is the basic raw material of any plant breeding programme. It is important that genetic diversity is maintained if crop development is to continue and that new characters are introduced into already existing cultivated genotypic background.

Why is genetic variability so important? Well it has been continually stated that without genetic variability, there can be no gain from selection. A further need is related to the appearance of new forms of pest or disease or new husbandry techniques, or new environmental challenges. If a new disease became important in an agricultural area to which all known cultivars were susceptible then it may be possible to identify new sources of disease resistance from closely related wild or weedy species.

There is a growing awareness of reduced germplasm resource throughout modern agriculture. The greater use of mono-culture crops and homozygous cultivars has greatly reduced the genetic variability within our agricultural crop species. For example, at the turn of the century, farmers growing cereal crops were propagating land races which were a collection of genetically different types grown in mixture. Land races have been almost completely replaced, in most countries, by homozygous lines or hybrids and much of the variability that existed has already been lost. Disease epidemics can also greatly reduce genetic variability within a crop species. The potato blight which affected western Europe (not just Ireland) had the effect of greatly reducing the genetic variability within European potato lines. Modern agricultural has become heavily reliant on chemical weed control. In our agricultural systems weeds can be almost completely eliminated leaving only the single homozygous genotype that was planted by the farmer.

Worldwide organizations have been formed with the specific aim of conserving germplasm which is accessible to breeders to search for new traits that are not available within the cultivated crops. The International Plant Genetic Resources Institute (IPGRI) is one organization which coordinates germplasm collection activities on an international level. IPGRI is part of the Consultative Group on International Agricultural Research (CGIAR).

In addition to the national germplasm collections and IPGRI, other organizations in the Consultative Group on International Agricultural Research (CGIAR) research centres, such as the International Potato Research Center (CIP, Peru), the International Center for Maize and Wheat Improvement (CIMMYT, Mexico), the International Rice Research Institute (IRRI) and the International Crops Research Institute

for the Semi-Arid Tropics (ICRISTAT, India) have remits to maintain germplasm collections on specific crop species.

Germplasm is available within the United States from the Plant Introduction System. Genotypes are made available from the location which maintains plant introduction material or from one of the regional stations. Some of the major crop responsibilities of each station are as follows:

- **Northeastern Regional Plant Introduction Station, Geneva, New York:** Perennial clover, onion, pea, broccoli and timothy
- **Southern Regional Plant Introduction Station, Georgia:** Cantaloupe, cowpea, millet, peanut, sorghum and pepper
- **North Central Regional Plant Introduction Station, Ames, Iowa:** Corn, sweet clover, beets, tomato and cucumber
- **Western Regional Plant Introduction Station, Pullman, Washington:** Alfalfa, bean, cabbage, fescue, wheat, grasses, lentils, lettuce, safflower and chickpea
- **State and Federal Inter-regional Potato Introduction Station (IR-1), Sturgeon Bay, Wisconsin:** Potato

Germplasm in itself is of little use to a plant breeder unless there is information regarding the attributes or defects of different genotypes. Most germplasm collections have associated data banks detailing and classifying material within the collection. For example, the Germplasm Resources Information Network (GRIN) is a computerized data base containing information on the location, characteristics and availability of accessions within the plant introduction scheme. This information is available to any breeder through the Database Management Unit of the Agricultural Research Service, Plant Genetics and Germplasm Institute, Beltsville, Maryland.

THINK QUESTIONS

(1) Selection in a plant breeding programme can be divided into three different stages: early generation selection; intermediate generation selection; and advanced generation selection. Briefly, state the major differences between the three above stages.

(2) A 10 × 10 half diallel crossing design was used to examine the potential of each of ten parents in

a wheat breeding programme. F_1 families along with each of the parents were grown in a properly designed field trial. Yield (kg per plot) was recorded and from the data, array means, within array variances (V_rs) and between array covariances (W_rs) were estimated. These statistics are shown below. From this information determine which three of the ten potential parents would be best suited for use in a cultivar development scheme designed to increase wheat yield? Explain your choices.

Parent	Mean	V_r	W_r
1	32.3	234.0	215.2
2	15.2	45.2	19.2
3	21.3	150.4	298.1
4	24.5	17.3	19.2
5	29.3	100.1	90.1
6	17.4	210.9	250.3
7	16.3	199.0	99.1
8	19.1	26.9	15.6
9	17.1	292.8	211.2
10	22.3	379.5	403.1

(3) In a bean breeding programme, it is desired to produce new cultivars which are short (dwarf) in stature and that have oval bean shapes (rather than round). Both bean shape and dwarfism are controlled by single alleles at one locus. The allele for oval beans is dominant to the alternative allele for round beans. The dwarfing gene is recessive to the non-dwarf (tall) gene.

A cross is made between two homozygous parents where parent 1 is dwarf and with round beans (*ttrr*) while parent two is tall with oval beans (*TTRR*). A number of F_1 plants are selfed to produce F_2 seeds from which 1600 F_2 plants are grown. Assuming independent assortment of genes, outline a selection scheme which will result in harvesting F_4 seeds that are homozygous for oval beans and dwarf stature. Indicate the number of plants selected at each selection stage.

(4) You are a potato breeder working in a publicly funded organization. Due to the break-up of the former Soviet Union, you have inherited 500 potato lines from the Siberian Potato Research Center. Briefly outline (using diagrams if necessary) how you would screen these genotypes for their potential as new parents in your breeding programme.

(5) In a winter rapeseed breeding programme 3000 near-homozygous breeding lines were evaluated for yield and oil content from a properly designed field trial. The correlation coefficient between yield and oil content was found to be $r = 0.41$. Using independent culling you want to select the highest yielding 10 per cent and select for oil content retaining the best 15 per cent. How many genotypes would you expect to be: (1) selected for high yield and high oil content; and (2) selected for high yield but discarded for high oil content?

(6) Independent culling can be a very effective means to select for two or more characters simultaneously and can be easily applied to breeding data. Under what circumstances would independent culling not be very effective? If index selection is used list two methods that could be used to weight variates.

(7) 50 progeny from ten potato crosses were evaluated for yield in a properly designed field trial. From the results the following cross means and genetic variances (σ_g^2) were obtained:

	Mean	σ_g^2
Cross 1	25.60	27.34
Cross 2	19.33	19.40
Cross 3	27.71	13.31
Cross 4	12.06	10.39
Cross 5	13.11	15.63
Cross 6	26.56	14.21
Cross 7	27.45	25.69
Cross 8	19.21	15.21
Cross 9	23.21	39.13
Cross 10	19.32	17.31

Also grown in the same trial were ten commercial cultivars. The average performance of the commercial cultivars was 20.14 and the standard deviation was 3.26. Using univariate cross prediction procedures determine which three crosses should be used in breeding for cultivars which would have high yields. Rank your choices as first, second and third and include the probabilities used for your decision.

- Greater than average controls plus one control standard deviation
- Greater than the average control plus twice the control standard deviation

Using the data presented above, how many clonal lines would need to be raised from Cross 6 to be 90% certain of having one line which would have a yield potential exceeding 3 standard deviations from the control mean.

(8) A half diallel crossing design is carried out involving 30 homozygous parents. F_3 progeny from each of the 435 possible cross combinations were raised and grown in a two replicate yield trial (planting F_3 and harvesting F_4 seed). Also grown within the trial were all 30 parental lines. Based on the yield results from the trial it was found that the regression of mid-parent value (x) onto progeny performance (y) was:

$$y = 0.832\,x + 0.002$$

Also from this trial, the phenotypic variance of the F_3 families was found to be $\sigma_p^2 = 65.216$. What would be the expected gain from selection if the F_3 families were selected at the 10% level (i.e. discard 90% of families) according to yield performance? Would you expect the same response to selection if the 43 selected F_4 families were further selected for yield the following year? (Explain your answer).

(9) In chickpea breeding it is known that the correlation between yield performance of F_4 families in one year and F_5 families in the following year is $r = 0.57$. 4000 F_4 breeding lines of chickpea were evaluated for seed yield in a properly designed field trial. All 4000 lines were re-evaluated at the F_5 stage the following year. If the highest yielding 10% were selected based on F_4 performance and the highest yielding 15% were selected based on their F_5 performance, how many of the original 4000 lines would you expect to be selected both on F_4 and F_5 performance. Explain what Type I and Type II error means in the context of selection. Estimate the Type I and Type II errors expected by selecting F_4 families for yield at the 10% level and selecting F_5 families at the 15% level.

(10) Parental selection can generally be divided into two different types. List the types and, briefly, indicate differences between them.

(11) Describe the main features of North Carolina I, North Carolina II and Diallel crossing designs. Explain the terms in the model for the analysis of a diallel according to the method described by Griffing:

$$Y_{ijk} = \mu + g_i + g_j + s_{ij} + e_{ijk}$$

and indicate the importance of these terms and Griffing Analysis in selecting superior parental lines.

(12) You have been appointed as Assistant Professor/Plant Breeder in the Crops Division of McDonalds University in Frysville, MD. It appears that the breeding programme has been trying to select improved genotypes with decreased sugar content in the tubers. However, in the 10 years previous to your appointment, there appears to have been no genetic improvement resulting from breeding. Outline three reasons that could individually, or in combination, have caused this non-response.

(13) In a series of properly designed experiments, a team of plant breeders produced estimates of narrow-sense heritabilities (h_n^2) for plant height and plant yield from two different segregating families of spring barley (95.BAR.31 and 95.BAR.69). Both families were grown on two farms (Moscow and Boise, in Idaho) in three successive years (1993, 1994 and 1995). The h_n^2 values from each year and site are summarized below.

Character	Year	95. BAR. 31		95. BAR. 69	
		Moscow	Boise	Moscow	Boise
Plant	1993	0.20	0.22	0.83	0.86
height	1994	0.01	0.21	0.31	0.74
	1995	0.21	0.30	0.52	0.79
Plant	1993	0.11	0.25	0.52	0.57
yield	1994	0.02	0.15	0.12	0.46
	1995	0.21	0.24	0.21	0.48

Which family is likely to give better responses in a breeding programme, given equivalent selection intensities, and why? In such a breeding programme, which of the two characters (plant height or plant yield) is likely to give the better response to equivalent selection intensities, and why? Consistently higher average values of h_n^2 were obtained at the farm near Boise. Which site, if either, is likely to provide the most accurate estimate of h_n^2, and why? Some of the heritability estimates (particularly those for plant height for 95.BAR.69 in Moscow) varied greatly over the 3 year period. What could be the cause?

(14) 4000 F_3 lentil breeding lines were evaluated for yield and the highest yielding 500 genotypes selected. The average yield of the 500 selection was 1429 kg/ha while the average yield of the discard genotypes was 1204 kg/ha. A random sample from the discards and the 500 selected lines were grown in a properly designed F_4 trial where the average yield of the random genotypes was 1199 kg/ha and the yield of the selected 500 lines was 1362 kg/ha. Determine the narrow-sense heritability for yield at the F_3 in lentil.

(15) In a high diastase barley breeding programme it is known that the correlation between yield and diastatic power is $r = 0.60$. In this breeding programme 3000 doubled haploids are evaluated for yield and diastatic power in a properly designed field trial. If you wish to retain approximately the 'best' 150 lines what proportion would your discard based on yield and what proportion would you discard based on diastase to achieve this selected number.

(16) In 1998, 10 000 F_3 head-rows were grown in an unreplicated, but randomized, design from *Bobby Z's Wheat Breeding Program*. At harvest, seed from all rows were thrashed and weighed. The average yield of all 10000 lines was found to be 164.24 kg. After weighing, 200 rows were taken at random and retained irrespective of their yield performance. From the remaining lines the 'best' 1000 rows were selected for further evaluation in the breeding scheme. The average yield of these selected lines was of course higher than the whole population at 193.74 kg. In 1999, the bulk F_4 seed from the 200 random lines and the 1000 selected lines were grown in a randomized complete block design. At harvest, each plot was harvested separately and the yield recorded. From this 1995 trial it was found that the average yield of the selected lines was 168.11 kg, while the average yield of the

200 randomly chosen lines was 133.41 kg. Determine the narrow-sense heritability for yield using the above information. Explain how this value of heritability found might influence your selection strategy at future F_3 head-row selection stages.

(17) In a Douglas fir breeding programme you have only sufficient resources to screen segregants from two cross each year. However, in 1995, you have been provided data of family mean (MEAN), phenotypic variance (VAR), and narrow-sense heritability (h^2) from six crosses. From these three statistics determine which two crosses you will put your efforts in 1995 assuming that you will select at only the 10% level, and explain, briefly, your choices.

Cross code	MEAN	VAR	h^2
DF.33.111	22.3	16.7	0.45
DF.66.123	24.6	14.3	0.51
DF.97.37	28.1	6.3	0.47
DF.97.332	26.1	15.3	0.11
DF.99.1	22.5	10.2	0.84
DF.99.131	18.9	26.1	0.75

(18) 25 progeny from ten winter wheat crosses were evaluated for yield in a properly designed F_3 cross prediction study under field conditions. From the results the following cross means and additive genetic standard deviations (σ_A) for seed yields were obtained:

	Mean	σ_A
92.WW.46	236.60	127.34
92.WW.53	199.33	191.40
92.WW.54	241.82	91.43
92.WW.61	142.00	119.10
92.WW.71	133.11	125.46
92.WW.74	236.73	102.14
92.WW.93	233.55	281.77
92.WW.108	201.22	106.63
92.WW.111	229.37	299.39
92.WW.116	169.11	119.32

Also grown in the same trial were five commercial cultivars. The average performance of the commercial cultivars was 211.10 and the standard deviation was 41.93. Using univariate cross prediction procedures determine which three crosses should be used in breeding for cultivars which would need to produce yields.

- Greater than average controls plus **one** control standard deviation
- Greater than the average control plus **twice** the control standard deviation

Rank your choices as first, second and third and include the probabilities used for your decision.

(19) 500 dry pea F_6 breeding lines were evaluated for yield potential at two locations (Hillside and Nethertown). After harvest and weighing, the top 15% highest yielding lines were selected at each location. When results from this selection were examined it was found that 39 lines were selected in the top 15% at each site. From this information, estimate the narrow-sense heritability.

In this same study, the following phenotypic variances and site means yields (over all 500 F_6 lines) were:

	Hillside	Nethertown
Phenotypic variance	96 kg^2	124 kg^2
Site mean	27 kg	29 kg

Given these data and the heritability estimated above, determine the expected response from selection at 10% level.

(20) Describe the difference between general combining ability and specific combining ability.

The following are average plant heights (over four replicates) of a 5×5 half diallel (including selfs) between sweet cherry cultivars.

Golden Glory	112				
Early Crimson	72	53			
Sweet Delight	102	64	99		
Dwarf Evens	56	41	65	49	
Giant Red	130	100	109	107	115
	Golden Glory	Early Crimson	Sweet Delight	Dwarf Evens	Giant Red

Estimate the general combing ability of each parent and calculate the specific combining ability of each cross. Which parents would be 'best' in a breeding programme to develop short cultivars?

(21) What is the difference between a cultivar with *general environmental adaptability* and one with *specific environmental adaptability*?

Eight yellow mustard F_8 breeding lines were grown at 20 locations throughout the Pacific Northwest region. Significant genotype $\times$ environment interaction was detected for yield by analysis of variance, and a joint regression analysis carried out. The following are line means and environmental sensitivity $(1+\beta)$ values from the analysis.

Line	Overall mean	Environmental sensitivity
90.EW.34.5	1234	0.55
90.JB.456	1890	1.05
90.JB.562	2345	1.34
91.HG.12	1897	0.76
91.HH.145	1976	0.52
92.22.12	2567	1.42
92.AE.1	2156	0.83
92.HK.134	2152	1.72

Select the two '*best*' breeding lines which show general environmental adaptability. Select the two '*best*' breeding lines which show specific environmental adaptability.

8

Alternative Techniques in Plant Breeding

INTRODUCTION

In this book so far it has been assumed that plant breeding involves artificial hybridization between chosen parents and selection (using visual assessment or by means of recording data) of desirable recombinants, over several generations. However, there is a range of techniques available to plant breeders that have made, or are starting to make, contributions to the production of new cultivars. These include: induced mutation; interspecific species hybridization; *in vitro* propagation; and plant transformation, all of which have been used to increase the genetic variability available to breeders. In addition, markers (mainly molecular based markers) are being increasingly used in breeding program, for example to aid selection for characters that are difficult to evaluate in the usual way.

INDUCED MUTATION

The variation that exists within all living plants and animals, including all crop species, is the result of natural mutations at the DNA level, with subsequent recombination and selection occurring, much of it over millions of years. But this has also been accompanied by changes at a structural level, such as rearrangement within and between chromosomes.

Mutations result in the generation of additional genetic variation within plant species. It has been estimated that mutations occur naturally with a frequency of 1 in 1 000 000 (one in a million) individuals. Most of these mutations are recessive and deleterious and these new alleles usually do not survive at anything other than

very low frequency in nature. However, if the mutation results in an advantageous effect, the genotype possessing the mutation may thus be more adapted to the environment compared to the non-mutant types and hence will tend to leave more offspring. Over generations this therefore leads to an increase in the frequency of the new allele within the population.

Obviously, the extremely low rate of natural mutation and even lower frequency of *desirable mutation* events are such that natural mutation has had little impact on modern plant breeding. However, in the mid 1920's it was discovered that X-rays could be used to induce high mutation rates first in fruit fly and later in barley. Plant breeders were quick to realize the potential of induced mutation and mutation breeding became a common practice in almost all crop species and in many ornamental flower breeding programmes.

The aim of mutation breeding is to stimulate an increase in the frequency of mutation events within crop species and then to select desirable new alleles from amongst the mutants produced. More basically, mutation breeding has been utilized to make minor advantageous genetic changes in already established and adapted cultivars through induced mutation treatments. For example, by inducing mutation in a highly adapted crop cultivar, and screen the resulting mutated lines for a specific character of interest. In doing this it is hoped to retain the existing cultivar adaptability while adding the mutated *advantageous* trait.

Mutagenesis derived lines in a plant breeding scheme are labelled according to the number of generations after mutation has taken place. For example, the generation immediately after mutation is termed the M_1. These plants can be self pollinated to produce an M_2 generation, and so on (compared with F_1, F_2, F_3, etc. for the more usual sexual generations).

Method of increasing the frequency of mutation

In general terms there are two methods that have been used to produce an increased frequency of mutations in plant species, **radiation** and **chemical** induction, with the highest frequency of mutation derived cultivars being from radiation induced mutants.

Mutations following exposure to radiation are produced by a variety of effects from physical damage through to disturbing chemical bonds. Two main types of radiation have been utilized to induce mutation in crop breeding schemes. These are:

- **Gamma rays** are the most favoured radiation source in plant breeding and have been used to develop 64% of the radiation-induced mutant derived cultivars. Gamma rays represent electromagnetic radiation with a high energy level and are produced by the disintegration of radioisotopes. The two main sources of gamma radiation for induced mutation are from cobalt 60 and cesium 137. Plant breeding programs have used gamma ray radiation treatments applied in a single dose or have treated whole plants to long exposure to gamma radiation.
- **X-rays** were the original radiation mutagen, yet have been responsible for only 22% of the cultivars released world wide from mutation-induced breeding programs. X-rays are produced when high-speed electrons strike a metallic target. X-rays are high energy ionizing radiations which have wavelength ranging from ultraviolet to gamma radiation. Mutations are induced by exposing seeds, whole plants, plant organs, or plant parts to a source X-ray radiation of a required frequency for a specific time. X-rays have to be handled by trained radiologists, and they are not always easily accessible to plant breeders, who often have to rely on medical facilities (i.e. hospitals) for mutagenic treatment of plants.

Other forms of radiation that has been used to induce mutation include **neutrons** (an electrically neutral elementary nuclear particle produced from nuclear fission by uranium 235 in an atomic reactor), **beta radiation** (negatively charged particles that are emitted from radioisotopes such as phosphorus 32 and carbon 14) and **ultraviolet radiation** (used primarily for induced mutation in pollen grains).

Many of the mutagenic chemicals are alkylating agents and include: sulphur mustards; nitrogen mustards; epoxides; ethylene-imines; sulphates and sulphones; diazoalkanes and nitroso-compounds. The most commonly used chemical mutagens have been ethyl-methane-sulphonate (EMS) and ethylene-imine (EI). It should be noted that all mutagenic chemicals are highly toxic and highly carcinogenic.

The frequency of mutation can also be influenced by oxygen level in plant material (higher oxygen related to increased plant injury and chromosomal abnormalities), water content (also related to oxygen content) and temperature.

Types of mutation

Mutations can be conveniently classified into four types:

- **Genome mutation** where there are changes in chromosome number due to either addition or loss of whole chromosomes or sets of chromosomes.
- **Structural changes in the chromosomes** involving translocation (a chromosomal aberration involving an interchange between different non-homologous chromosomes), inversions (changes in the arrangement of the loci, but not in their number), deficiencies, deletions, duplications and fusions (reduction or increase in the number of loci borne by the chromosome).
- **Gene mutations**, often termed point mutations, where the change is in a single gene, and often the result of a single base pair change at the DNA level.
- **Extra nuclear mutations** where the mutational event occurs in one of the cytoplasmic organelles. The DNA involved includes plastids and mitochondria and means that the mutation will usually be transmitted from one generation to the next through just one of the sexes, usually via the egg cells, i.e. maternally inherited (one example of this form of mutation is cytoplasmic male sterility, common in many crop species).

Plant parts to be treated

Mutagenic agents can be applied to different parts of plants and still produce effects. Seeds can be treated with either chemical or radiation mutagens. Seed have been preferred by many breeders because seeds are more tolerant to a wide range of physical conditions such as being desiccated, soaked in liquid, heated, frozen or maintained under varying oxygen levels, which allows their exposure to the various mutagens rather easily. Seed treated with an induced mutagen will result in plants which are:

- Non-mutants, the same as the parent plant.
- True mutants, where a mutation event has occurred throughout the whole plant.
- Chimera, where only a portion of the resulting plant has been mutated.

One attractive possibility in mutation breeding is to treat pollen grains with radiation or chemical mutagens. The major advantage of treating pollen grains is that they are easily collected in large numbers and can easily be presented to a radiation source. Pollen grains are single cells, so induced mutation of pollen avoids the occurrence of chimeras. Pollen grains are also haploid in terms of their genetic composition and so this opens the possibility, in an increasing number of species, of tissue culture treatment leading to their direct development into plantlets – which, with suitable treatment, can be induced to double their chromosome number and give true breeding, homozygous lines which will express both recessive and dominant mutated alleles.

Treatment of whole plants is less common but can be achieved using X-rays or gamma rays. This is often carried out using small plants or plantlets but, for example in Japan, they have built a large facility (resembling a sports arena) with a large gamma source at its centre and a large number of fully grown plants are exposed over varying periods.

The treatment of cuttings and apical buds with radiation or chemical mutagens can be effective in developing mutant types in new shoots and plantlets. An important factor is whether the meristematic region forms mutations, since this is the region from which the new propagules develop. Treating cuttings and apical buds has been particularly important in developing mutant clonally propagated cultivars.

It is now becoming more popular to combine mutation with *in vitro* cell and plant growth. The idea centres on mimicking that possible with microorganisms and so often involves treating single cells with a chemical or radiation mutagen and screening regenerating plantlets. One very desirable approach is the use of selective media, which only allows the growth of specific mutant types. This has been useful in developing herbicide resistant cultivars where a low concentration of the selected herbicide is added to the media.

Dose rates

It is apparent that all mutagenic treatments are basically damaging to plants. When too high a dose rate is applied all the plant cells may be killed. Conversely if the applied mutagen dose rate is too low, then very few mutant types will be induced. It is therefore necessary to determine an appropriate dose rate to use. The optimal dose rate will change according to crop species, plant part exposed to the mutagen and its physiological state. Indeed the first stage of most mutagenesis based breeding is to determine the most appropriate dose rate to minimize adverse effects, yet still produce sufficiently high levels of mutation. In practice it is usual to carry out several preliminary experiments in order to establish a suitable dose rate and protocols, which optimize the survival of the plant material subsequent to exposure.

In simple terms dose rate is equal to mutagenic intensity × time applied. In chemical mutagenesis this involves the concentration of mutagenic chemical and time that plant cells were exposed to the chemical solution. Intensity of radiation can be altered by varying the distance from the radiation source or by varying the radiation form. The dose can be adjusted by changing mutagenic intensity or exposure time (or both). It is common to experiment with different dose treatments until one is found which allows 50% of plants to survive the treatment. These tests are called **lethal dose 50** or LD_{50} tests.

Dangers of using mutagens

Mutagenic chemical and radiation are effective because they alter the genetic makeup of plants and create variation. They will, of course, similarly, affect the DNA of plant breeders who are exposed to them! It is not therefore possible to over emphasis the importance of using appropriate safety procedures in using any mutagen. As already mentioned, the facilities for applying mutagenic treatments (in this case mainly radiation) are not always directly available to the average plant breeders, specialized operators or personnel (i.e. hospital radiologists, etc.) usually carry out the actual exposure to the radiation.

To use chemical mutagenic agents safely requires a number of safety features, spelt out in many countries (and by most suppliers in safety/hazard assessments) by specific safety protocols. Staff using these chemicals should be aware of the advised risks and safety procedures. Minimum safety will likely require suitable gloves, protective clothing and safety glasses combined with '*Good Laboratory Management Practice*'. It is also important that procedures and equipment are in place to deal with appropriate disposal of chemicals, and to contain and clean up any accidental spills of mutagenic chemicals.

Impact of mutation breeding

Mutation derived cultivars have been released as a direct result of mutagenesis or have used mutant genotypes as parents in traditional breeding programmes. Since the inception of mutation breeding over 2250 cultivars have been released world wide (FAO/IAEA [Food and Agricultural Organization/International Atomic Energy Agency] Mutant Varieties Database). It should be noted, however that a high proportion of mutation-derived cultivars released were ornamental plants and flowers rather than agricultural crops. Over 70% of these cultivar releases were developed directly from mutant breeding lines. Most of these cultivars were developed and released in Asia with 27% being developed in China and 11% developed in India. Mutation-induced cultivars are not quite as common in other countries although over 125 mutant induced cultivars have been released in the US and 32 in the UK (31 of which were barley cultivars) in the past 70 years.

Highest mutant cultivar releases were in rice (433), followed by barley (269), wheat (220), soybean (89), groundnut (47), maize (32), pea (32), cotton (24) and millet (24). Only 46 mutant fruit cultivars were released over this period. Mutant genes were developed mainly for dwarf stature, improved disease resistance, stress resistance, herbicide resistance, and improved grain or oil quality.

Although many have argued that these released cultivars have made little impact on our agricultural crops, some major positive impacts cannot be denied.

Semi-dwarf rice derived from mutation breeding has been cultivated over millions of hectares. The barley cultivar 'Diamant', developed as a gamma ray mutant of 'Valticky', was selected to have the *ert* dwarfing gene. It has been estimated that over 150 cultivars have been released in Europe that have Diamant in their pedigree. In addition the Scottish barley cultivar 'Golden Promise' also has this mutant dwarfing gene, is arguably the best cultivar ever released in the country and has been a major contributor to the Scottish brewing industry. Similarly, mutant durum wheat occupied over 25% of the Italian wheat acreage in the mid-1980s. Finally, health concerns about '*trans* fats' in our diets has prompted many food processors and others in the food industry to use non-hydrogenated vegetable fats which are low in polyunsaturated fats. The first canola cultivar with low linolenic acid, 'Stellar', inherited the fatty acid desyntheses gene from a German EMS-induced mutant line coded as M47. Other low linoleic acid mutants have subsequently been developed using microspore mutagenesis and now ultra-low polyunsaturated canola oil cultivars with highly elevated oleic acid content are commercially available.

Practical applications

Having decided which mutagenic agent to use and a suitable mutagenic treatment strategy (i.e. rate and time, and which plant part to treat) in reality, the physical treatment of plant cells by a mutagenic agent to induce mutation is in fact the easy part of mutation breeding. By far the most difficult aspect of mutation breeding relates to selecting desirable mutants while avoiding the subsequent detrimental effects of mutagenesis.

Mutagens are indiscriminating agents and inevitably produce a complex mixture of mutations. Mutants selected as having the trait of interest may have also undergone a range of chromosome structure changes and non-genetic (or at least non-nuclear) aberrations. It is also rare that selected mutants have been genetically altered for the single gene of interest, and there may be multiple mutation events, most of which have a negative impact on the normal growth of the plant. In seed-propagated crops sterile segregants need to be sorted out and discarded in the first round of selection. Therefore selected mutants will have been altered in number of different ways – most of them bad!

Consider the following example of using mutation breeding techniques in yellow mustard (*Sinapis alba*). The aim of the breeding programme was to screen a large number of mutants (derived by chemical mutagenesis using EMS) that produce lines that were low in seed meal glucosinolate content. After screening several thousand lines over a four year period (using glucose sensitive test tape) eight lines were identified which showed lower glucosinolates compared to the non-mutated parent genotype ('Tilney').

However when these eight lines were grown in replicated field trials it was found that all mutant lines were considerably lower in yield, produced smaller and less vigorous plants and matured later than the parent genotype (Table 8.1). So exposure to the mutagen had produced mutations that had affected glucosinolate content in the manner hoped for. However, it had also affected other aspects of the genotype so that the selected lines all appeared to have mutated for other important traits. Further crossing and selection was clearly necessary before an adapted cultivar was developed.

It has to be pointed out that experience suggests that apart from loss of function alleles, it is usually easier to *find* a new allele than to *create* one. It may be significant to note that several years of intense mutagenesis breeding effort to develop yellow mustard lines with very low glucosinolate content in the seed meal failed to achieve this objective. Interestingly, the year after the mutagenesis program was stopped, a gene which almost eliminated seed glucosinolates in yellow mustard was identified within a wild *Sinapis alba* population from Poland!

Initially many plant breeders believed that mutation breeding would have a revolutionary effect on cultivar development. Although there are multiple examples, like those above, of success in mutation breeding, for many this revolution has not happened and is unlikely to now do so. Indeed, many believe that mutation breeding is too unpredictable and should only be considered as a *last resort technique* when all other avenues have closed. Whether one is a mutagenasist or not, it is apparent that mutagenesis as a plant breeding technique has its limitations.

The question must therefore be asked as to what are the circumstances in which mutation techniques can be useful? To address this question the following points should be considered.

- Mutagenesis is an indiscriminate breeding approach and generates large numbers of undesirable variants along with those that are wanted.
- Most successful mutant cultivars and mutation derived cultivars have resulted in selection for characters controlled by single (or at best a few) genes. Quantitatively inherited traits are more challenging in mutation breeding and will require large efforts to achieve success. If indeed success is ever achieved.

Table 8.1 Seed yield, oil content, ground cover, days from planting to flowering and days from planting to flower ending of the yellow mustard cultivar 'Tilney', and eight EMS mutants selected from Tilney with modified seed quality.

Identifier	Seed yield (kg/plot)	Oil content (%)	Percentage ground cover 1–9 scale	Days to flowering (days)	Days to flower end (days)
Tilney	27.33	30.1	8.7	54	80
Til.M3.A	7.00	30.5	3.0	59	86
Til.M3.B	22.67	29.2	4.7	58	85
Til.M3.C	12.67	30.4	2.0	60	86
Til.M3.I	17.00	29.7	4.7	59	86
Til.M3.II	15.67	29.7	3.0	60	87
Til.M3.III	9.67	30.1	2.3	59	86
Til.M3.IV	18.67	31.0	3.3	58	86
Til.M3.V	15.33	31.0	5.0	58	86

- The frequency of desirable mutants is likely to be low, and it is essential that there are suitable selection techniques to screen the thousands of mutants generated to identify the few that have the required mutated trait. The most common example of mass selection in mutation breeding relates to herbicide tolerance. Most herbicide tolerance in crop species is qualitatively controlled and many thousands of mutated breeding lines can easily be screened for tolerance by simply spraying the mutants with the herbicide of choice with the premise that all those that survive carry a form of the mutant tolerant gene. Other examples would include selecting mutants which exhibit morphological changes in plant structure (i.e. dwarfs), maturity, flower colour, qualitative disease resistance, or enhanced end-use quality (i.e. fatty acid profile or starch types).

- When desirable mutants are selected they are usually adversely affected by the mutagenic treatment and it will be necessary to 'clean up' the mutant genotype either by recurrent selfing and selection, or more likely by recurrent back-crossing and selection. Most mutations are recessive, and identification of recessive mutations is difficult due to dominance effects. In diploid seed-propagated crops, recessive mutations can be identified by selfing or inter-mating mutated lines, albeit that the frequency of lines that are homozygous recessive for the desirable gene mutation will be very low. For obvious reasons selection of desirable recessive mutations in clonally propagated crops is considerably more difficult and breeders would require excessive effort in selfing mutant lines or hope for the extremely rare event where all alleles at a given loci have the same recessive mutated gene.

The unpredictable nature of mutagenesis raises the question arises as to whether it is possible to 'direct' the mutational effects towards changing only characters of interest and to affect them in rather particular ways. The first possible 'direction' to the affects that can be exploited is that different mutagens have different forms of action, as noted earlier. Some induce point mutations and so are more likely to produce a particular array of effects while others are likely to induce grosser structural changes. An even more specific array of possibilities is arising from the potential to induce site specific mutagenesis. Consider our increasing ability to identify particular genes, to clone or synthesize these

and introduce them back into the plant species (or, of course, to another) – clearly the potential to 'mutate' these DNA sequences and reintroduce them is a reality.

One lesson, which was learned from the early efforts of mutation breeders, is that it is necessary to have clear objectives, which are biologically reasonable, if success is to be achieved. However, even in cases which have been well organized with realistic objectives, the effort that was required to sort out the desired products in a useable form was often greater than what would have been required to achieve the same results from a more traditional hybridization breeding scheme.

In plant breeding there will always be a need for new sources of variation and mutations (natural or induced) will feature as part of future breeding efforts. Therefore, mutation breeding has a very real place in cultivar development, but it would be unwise to base a complete variety development program on mutagenesis.

In summary, a mutagenesis breeding program must deal with large numbers of mutated lines so that the low frequency of desirable mutations, in an acceptable genetic background, can be selected. Similarly, a mutation scheme must offer a quick and effective selection screen to identify the few desirable mutants.

INTERSPECIFIC AND INTERGENERIC HYBRIDIZATION

Another method of increasing the genetic diversity of a crop species is by interspecific or intergeneric hybridization. When sources of variation for a character of interest (e.g. disease or pest resistance) cannot be found within existing genotypes in a species, it seems sensible to look at related species or genera and examine the possibility to introgress traits from them into the one of interest.

Interspecific hybridization refers to crosses between species within the same genus (i.e. *B. rapa* × *B. oleracea*) while intergeneric hybrids are crosses between different genera (*Triticum* × *Secale*).

The probability of developing a successful new cultivar is related to the frequency of desirable (or undesirable) characteristics in the parents used in hybridization. The most commonly used parental lines will be adapted cultivars or highly desirable genotypes from within breeding programmes. When a character of interest is not available within this gene pool then, the obvious next step is to screen other lines that are not as adapted, in an attempt to identify expression of the desired trait in them. If the character cannot be identified within this wider germplasm source then breeders spread their search wider and will screen related species in an attempt to find a natural genetic source.

Successful interspecific or intergeneric hybridization should therefore be considered when:

- The desired expression of a character of interest is not available within the gene pool of adapted genotypes, or their unadapted counterparts from the same species
- Acceptable expression for this trait has been shown to exist within a related species or genera
- It is possible to introgress alleles from the related species into the cultivated species

Successful interspecific crossing depends on two factors: obtaining viable seeds from plants in the F_1 (and later generations) and eliminating undesirable characters from the donor species. One, or both, of these factors may be the major determining factor in the actual success in gene transfer between species by this approach (see later for the possibilities using genetic transformation).

Characters introduced to crops from wild related species

A high proportion (over 80%) of genes introduced to our crop species through interspecific or intergeneric hybridization relate to pest and disease resistance. This trend continues today whereby wild related species to our crop species are continually being screened and evaluated to identify new genes for resistance to crop diseases. Resistance to grassy stunt virus was introgressed from *Oryza nivara* to cultivated rice. A number of late blight resistance genes have been transferred from *Solanum demissum* into potato cultivars. In addition most new potato cultivars released in the EU contain the H_1 gene conferring resistance to potato cyst nematode (*Globodera rostochiensis*) transferred from *S. verni*. Cabbage seedpod weevil resistance has been transferred to rapeseed through intergeneric hybridization between *Brassica napus* and *Sinapis alba*. More recently genes conferring resistance to Hessian fly, a major insect pest of wheat in the US, have been transferred from *Aegilops tauschii*.

Other traits that have been transferred through interspecific or intergeneric hybridization include abiotic stresses, drought tolerance, heat tolerance and salinity tolerance. Examples of enhanced yield or quality characters from wild relatives into crop plants, not surprisingly, are very rare.

Factors involved in interspecific or intergeneric hybridization

In order to make a successful hybrid involving two different species, there needs to be some degree of compatibility between the parents used. A number of factors need to be addressed to ensure successful gene introgression.

The first stage, which must be overcome, is that the male and female gametes from the different genotypes must unite to form a zygote. Failure at this stage can result from:

- Inability of pollen grains to germinate on the receptive stigma of the female parent
- Failure of pollen tubes to develop successfully and grow down the style or non-attraction of pollen tube towards the ovary
- Inability of male gametes that do reach the embryo sac to actually fuse with the egg cell
- Inability of the nuclei from pollen and egg to fuse

All of these aspects are related to fertilization barriers, and a number of techniques can be used to overcome incompatibility at each stage. *In vitro* fertilization (i.e. using excised organs) can sometimes be used to overcome some incompatibility factors involved in the first two barriers listed above.

Success in interspecific hybridization may be unidirectional (i.e. style length differences that cause pollen tubes to fail to reach ovary) and these can be overcome by attempting the reciprocal cross. Therefore successful hybrids might be possible from the mating A × B but difficult, or unsuccessful, when tried as B × A. A good example of this is seen in the cross *Brassica napus* × *B. oleracea* that will produce viable hybrid seed if the cross is carried out in this direction (i.e. *B. napus* as female). If, however, *B. oleracea* is the female, then very few or no seed is produced (without using tissue culture techniques).

Cross incompatibility resulting from failure of pollen grains to germinate and develop pollen tubes is associated with proteins on the pistil that interact unfavourably with proteins in the pollen. In some cases this reaction has been overcome by mixing pollen from the donor species with compatible pollen from the female species.

Pollen tubes often fail to reach the ovary (or 'miss' the ovary) due to the physical differences in style lengths between the different species. This can sometimes be overcome by mechanically reducing the style length of the longer style parent. Although this will only be successful if the shortened pistils remain receptive (i.e. as in maize). In the extreme case the complete pistil can be removed and pollen applied directly into the ovary, usually requiring *in vitro* techniques.

Once fertilization is achieved the problems are not necessarily over. When two species differ in ploidy level it may be necessary to reduce or increase the ploidy of parents prior to crossing. In potato, potato cyst nematode resistance was identified in *Solanum verni* (a close relative to cultivated potato, *S. tuberosum*). *S. verni* is a diploid while *S. tuberosum* is tetraploid.

Two methods of successful hybridization have been achieved:

- Doubling the ploidy level of *S. verni*, using colchicine and then carrying out interspecific hybridization at the tetraploid level
- Producing dihaploids from *S. tuberosum* (using parthenogenesis) and crossing the two species at the diploid level. Progeny from the hybrid cross are then doubled to the tetraploid level using colchicine or by spontaneous doubling resulting from callus growth *in vitro*

A similar manipulation of ploidy in interspecific crosses in potato was used to introgress late blight resistance (*Phytophthora infestans*) into cultivated potato cultivars. The source of blight resistance was found in a wild relative of cultivated potato (*S. demissum*), which is a hexaploid. A small proportion of tetraploid progeny can be obtained by crossing dihaploid *S. tuberosum* (see above and in haploid section) with the hexaploid *S. demissum*.

If attempts to obtain hybrid seed by means of sexual crossing fail, then somatic fusion (fusion of protoplast) may seem a realistic possibility. Genetic transfer

between two species may be feasible using protoplast fusion, followed by regeneration of plants from isolated wall-less cells (protoplasts). Resulting somatic hybrids will have the combined chromosome number of both parents (e.g. as in allotetraploids obtained by interspecific hybridization) so it may be necessary to first reduce the ploidy of parental lines or reduce the ploidy of hybrid combinations. However, the most difficult aspects of this technology are:

- Being able to regenerate plants from protoplasts, even without fusion
- Selecting fused heterokaryons from unfused or self-fused parental protoplast

After fertilization, failure of seeds to develop and/or to reach maturity can result from embryo and/or endosperm abortion or failure in the stages of embryo, or fruit development to complete their necessary stages to give mature seeds.

Successful fruit and flower retention after fertilization can be a simple function of a dependency on having a sufficient number of developing embryos. In some interspecific or intergeneric hybrids the number of fertilized ovules is too low to stimulate mature fruit development. Growth regulators (e.g. gibberellic acid) have been used as a means to encourage fruit retention. It has also been suggested that increasing the frequency of developing seeds in fruits (by applying a mixture of compatible and incompatible (mentor) pollen) can be used to avoid flower or fruit abscission.

Many interspecific or intergeneric hybridizations fail as a result of post-fertilization factors, which cause embryo or endosperm abortion. It may be possible to obtain hybrid plants despite abortion by using *in vitro* techniques such as:

- Ovule culture, where the complete ovary is removed from the plant and aseptically transferred to a growth media chosen where the nutrients are therefore provided and thus seed development proceeds
- Embryo rescue, where immature embryos are excised from the ovary and transferred to growth media; the chosen media therefore replaces the natural endosperm (which may have aborted or have been about to abort)

Sometimes a combination of both techniques is necessary to achieve interspecific hybrid seed. Early in the

Figure 8.1 Winter biennial forms of yellow mustard (*Sinapis alba*) (left) and Indian mustard (*Brassica juncea*) (right) produced through intergeneric hybridization.

embryo development the ovary is removed and cultured *in vitro* to achieve embryos, which are of a suitable size to successfully rescue and culture.

Finally, it is not uncommon in hybrid crosses that rather than resulting in hybrid combinations the resulting seed develops as matromorphic plants, which are thus derived from the maternal genotype. This characteristic has been developed to advantage in producing homozygous lines (i.e. *Hordeum vulgaris* × *H. bulbosum*). The seeds from interspecific crosses should thus be checked to ensure that the matromorphs are discarded if the desire is to produce hybrids – but retained if this feature is being used to produce haploids of the maternal genotype!

Hybrid sterility

In many cases the F_1 plants resulting from interspecific crosses are completely (or partially) sterile. A common technique used to overcome sterility, caused by lack of chromosome pairing, is to induce chromosome doubling in the hybrid, and hence develop alloploids. When doubled, it allows each chromosome to have a homologue with which to pair at meiosis, and thus reduce the infertility problem.

Backcrossing

After interspecific hybridization the resulting progeny will generally contain a large proportion of undesirable characters from the donor species, along with the character it was wished to introduce. In such circumstances it is necessary to carry out several rounds of backcrossing to the host species, with selection for the new character to obtain genotypes, which will have commercial worth. Any programme involving interspecific or intergeneric hybridization is therefore likely to be long-term.

Increasing genetic diversity

Many crop species have a relatively narrow genetic base and it is often advantageous to broaden genetic diversity by introgressing traits from related weedy species. Several crop species (i.e. rapeseed and wheat) have evolved as allopolyploids, whereby they contain complete chromosome sets from two or more diploid ancestors. Greater genetic diversity and variation can be achieved in breeding by resynthesizing the crop species from its ancient ancestors.

Creating new species

It is possible to create new crop species by intergeneric hybridization. Despite the possible attraction of this there are very few instances where new crops have resulted. Two notable examples include:

- **Triticale**, which resulted by intergeneric hybridization between wheat (*Triticum*) and rye (*Secale*)

- **Raphanobrassica**, which resulted from the intergeneric cross between kale (*Brassica oleracea*) and radish (*Raphanus sativus*)

When each of these new species was created there was great hope that they would have almost immediate high potential commercial value. However, in neither instance has this full commercialization occurred – at least not yet.

TISSUE CULTURE

A variety of techniques have been developed under the title of tissue culture. It is not the intention to cover the details of these techniques but to briefly consider a couple of them, enough to be able to give an idea of their application.

Haploidy

Establishing true breeding, homozygous lines (as noted earlier), is an essential part of developing new cultivars in many crop species. These homozygous lines are used either as cultivars in their own right or as parents in hybrid variety development. Traditionally, plant breeders have used the process of selfing or mating between close relatives to achieve homozygosity, a process that is time consuming. Therefore the opportunity to produce plants from gametic, haploid cells has been the goal of many plant breeders as this technique would produce 'instant' inbred lines once the chromosomes of the haploids are doubled.

The genetic phenomenon critical to obtaining homozygous lines is the formation of haploid gametes by meiosis. During this type of cell division, the chromosome number is halved and each chromosome is represented only once in each cell (assuming the species is basically a diploid one). If such gametic, haploid cells can be induced to develop into plantlets (i.e. we encourage the development of the sporophyte – *note*: lower plants often have this as a specific phase of the life-cycle) a haploid plant can develop which can then be treated (usually with a chemical called colchicine) to encourage its chromosomes to double, to produce a completely homozygous line (a doubled haploid).

Techniques used for producing haploids in vitro

Although haploidy is a very attractive technique to many plant breeders the natural occurrence of haploid plants is rare. However, the use of plant tissue culture has allowed the production of plants from gametic cells cultured *in vitro*.

Although haploid plants can be regenerated from both male and female sex cells, it is generally the male cells (microspores or pollen) that have proven most successful in the regeneration of large numbers of haploid and doubled haploid lines. This is partly because of the ease with which pollen, as opposed to eggs can be collected, and partly because it is simply that, in general, many more pollen grains than eggs are produced.

There are, of course exceptions and some examples include:

- The relative ease by which haploid barley plants can be produced from female sex cells. Interspecific crosses between cultivated barley (*Hordeum vulgare*) and the wild species *H. bulbosum* followed by *in vitro* culture of rescued immature embryos results in haploid plants as a result of exclusion of the *H. bulbosum* chromosomes during embryo development.

- Dihaploids from tetraploid potatoes have been produced in large quantity, using interspecific hybridization between cultivated potato (*Solanum tuberosum*) and a diploid relative (*S. phureja*). The cross of the tetraploid female *S. tuberosum* with the diploid male *S. phureja* would be expected to produce only triploid offspring – but it does not. Instead, the numbers of seeds obtained are relatively few and are predominantly tetraploid (as a result of the production of unreduced ($2n$) pollen from *S. phureja*). Among the rest are some of the expected sterile triploids but also some maternal dihaploids arising from the egg. Lines of *S. phureja* have been selected which produce a high frequency of dihaploid seed, greater than 70%. In addition such pollen parents have been selected to include a homozygous dominant embryo spot marker, which makes visual identification of the non-dihaploid seed easy.

There are other haploid induction mechanisms but the most widely applicable are via anther or microspore (immature pollen grains) *in vitro*. The anthers, of course, are flower organs in which microspores mature into pollen grains under normal conditions (i.e. *in vivo*). The production of haploid plants from anther culture has been reported for over 200 species of higher plants. However, although the technique offers great potential for use in plant breeding programmes the current examples of its application on a large, practical, scale are restricted, but some are provided by commercial programmes in: rice; wheat; barley; rye; rapeseed; tobacco; potato; pepper and maize.

Some potential problems

Genotype dependence

One factor, which has limited the use of anther culture in practical plant breeding programmes, is that even the different variants of the protocol often show strong genotypic dependence. Therefore, if a protocol is identified which is effective for one genotype that protocol often needs to be modified (sometimes to a large extent) to obtain success with another genotype or, more appropriately, a range of genotypes.

Somaclonal variation

The techniques noted above involve producing plants that have been regenerated following *in vitro* culture. Variation can often be detected among such plants that are regenerated and this variation has been termed somaclonal variation. The frequency of such variation has been suggested as reflecting the occurrence and length of the callus phase. In a haploid production scheme it is therefore essential that callus stages are kept to a minimum so that any somatic variation is kept to a minimum.

Non-random recovery of haploid lines

An important feature underlying the application of haploids in a plant breeding context is that the population of homozygous lines, derived from the chromosome doubling of the haploids produced, are a random representative of the gametic array possible. In other words, the possibility of unconscious selection occurring (effectively gametic selection) must be avoided. The genetic combinations recovered from haploid systems may be disproportionately composed of combinations from one of the original parents that were used to make the hybrid crosses from which the anthers were taken. An obvious possibility is that one

of the parents showed a much greater propensity for regeneration in culture and this would result in combinations with the genes that determined its response, being represented more frequently in the population of gametes. In experimental studies it has been shown that non-randomness of the possible gametic combinations can occur and can be influenced by the culture protocols used.

Practical applications of haploids

Progress in evaluating gametic-derived plants under field conditions has been increasing dramatically; reports using numerous crops have indicated the importance of continued research in this area.

However, it has been suggested that developing haploids in a practical breeding scheme will not be as effective as might be expected. In particular concerns have been raised regarding:

- The cost of producing haploids
- The inability to easily produce large numbers of homozygous lines through haploidy
- The deleterious variation that is sometimes exposed as a result of deleterious recessive alleles in the original material or mutational/somaclonal variation induced as a result of the *in vitro* techniques
- The dependence on the genotype of the parental material used in influencing the frequency of haploids produced – which often means that the very material the breeder most wants to use is non-responsive and haploids are not easily obtained

However, as refinements are made in methods and protocols, it is likely that it will become easier and cheaper to produce haploids on a routine basis. This will then mean that their impact on plant breeding programmes will be larger in the future. To date very few cultivars have been introduced as a direct result of haploidy (perhaps China being the exception). However there is little doubt that these techniques have added valuable information for plant breeders with regard to a number of aspects of genetics and tissue culture.

As noted earlier, one limitation to the widespread use of doubled haploids among many crops is the inability to produce large enough numbers of plants from culture. Regeneration frequencies are improving continuously, however, which will not only improve the applicability of the technique in a range of species but will also increase the potential for their application in other ways. For example, the possibility of deliberately applying positive selection pressure during the culture phase for certain characteristics, that is *in vitro* selection, will become even more attractive. Also this might be combined with induced mutagenesis during microsporogenesis, for example allowing production of novel resistance to fungal or bacterial pathogens or to herbicides.

In vitro multiplication

In vitro multiplication of breeding lines can have two main benefits (particularly in clonal species) in relation to plant breeding programmes:

- Plants propagated *in vitro* can generally be initiated to be disease-free, and can: be used to help maintain stocks of breeding lines; facilitate long-term germplasm storage and facilitate international exchange of material
- Short 'generation' times and fast growth means that rapid increases in plant number can readily be achieved

Both the above have particular importance to clonal crops which tend to have a relatively low multiplication rate as a result of their vegetative mode of propagation and which are particularly susceptible to viral and bacterial diseases, which tend to be multiplied and transmitted through each clonal generation.

Good examples of maintaining high disease status and offering rapid plant regeneration potential include potato and strawberry. Other, perhaps less well developed examples include *in vitro* propagation of date and oil palms. In these crops it was found that rapid plant regeneration would indeed offer an alternative to the slow and lengthy process of propagating side shoots in date palm and a more uniform planting material in the case of oil palm. However, in date palm the process is still very genotype dependent and with oil palm there proved to be an unacceptably high frequency of sterile palms produced with initial protocols, however,

these are now being revised and would appear to offer practical possibilities.

PLANT TRANSFORMATION

The stable introduction of foreign genes into plants represents one of the most significant developments affecting the production of crop species in a continuum of advances in agricultural technology relating to plant breeding. The progress in this area has depended largely on the tissue culture systems having been developed which, at least, initially, provide an amenable vehicle for the transformation induction.

The term transformation comes from that used for a much longer period, bacterial transformation, in which DNA has been successfully transferred from one isolate to another or between species of bacteria, and integrated into the genome. It was shown that the stably transformed bacteria then expressed the new genes and had appropriately altered phenotypes. In eukaryotes, transformation has a further complicating dimension, at least in many plant breeding contexts. The transforming DNA must not only be integrated into a chromosome, it must be a chromosome of a cell, or cells that will develop into germ-line cells. Otherwise the 'transformation' will not be passed on to any sexual progeny.

Using plant transformation techniques it is possible to transfer single genes (i.e. simply inherited traits) into plants, have such transgenes expressed and to function successfully. Theoretically at least, specific genes can be transformed from any source into developed cultivars or advanced breeding lines in a single step. Plant transformation, therefore, allows plant breeders to bypass barriers, which limit sexual gene transfer and exchange genes (and traits) from unrelated species where incompatibility does not allow sexual hybridization. These recombinant DNA techniques therefore apparently allow breeders to transfer genes between completely unrelated organisms. For example, bacterial genes can be transferred and expressed in plants. This therefore appears to break the barrier that sexual reproduction generally imposes. However, as we learn more and more about the DNA and hence the genes involved, the perspective of the picture somewhat changes. Increasing direct evidence of the presence in different species of the same basic gene, or clear variants of it, demonstrate the greater conservation of genetic material during evolution than we expected. Also, we are being reminded of the existence of parallel natural processes for much of what we regard as novel. For example, bacteria, viruses and phages already have successfully evolved mechanisms to transfer genes just in the way we regard as being so alien! But clearly, the new techniques are allowing modern plant breeders to create new variability beyond that existing in the currently available germplasm on a different scale and in a different time frame from that which was possible previously.

Although plant transformation has added (and some say dramatically) to the tools available to the breeder for genetic manipulation, it does (as with all techniques) have limitations. Some of the limitations will reduce with increased development of methodologies, others are those that are inherent to the basic approach.

At present, recombinant DNA techniques can generally only transfer rather limited lengths of DNA and so tend to be restricted to the transfer of single genes. This means that they are very effective where the trait can be substantially affected by one, or a few, gene(s) of large effect, but will be dependent on how much of the variation that is important in many agronomic traits showing continuous variation is actually controlled by a few loci showing rather large effects and how much by a myriad of ones with much lesser effects. So, for example, it is not clear how much yield itself, which could be argued is one of the most important characters of interest, can be manipulated by discrete steps of individual transformation events. Interestingly, however, recent reports do indicate the potential to transform with a number of genes (constructs) in one go with a reasonably high level of co-transformation.

It may seem obvious, but another restriction currently that is imposed, is that the techniques are only readily applied to genes that have been identified and cloned. The number of such desirable genes is still modest, but increasing rapidly. What is becoming clear is a deficiency in the knowledge of the underlying biochemistry or physiology of most traits. Another feature, which has recently provided at least a temporary limit to the technique, has been the identification of suitable promoters for the genes that are to be introduced. The inappropriate expression of a transgene, in the development of the plant, or particular organ, or in its timing has now been fully recognized and so the search for promoters now equals that for the genes themselves.

In addition, it has been recognized that because of the uncontrolled nature of the incorporation of transgenes into the host's genome a large number of transformed plants need to be produced in order to allow the selection of the few that have the desired expression of the transgene without any detrimental alteration of all the characters of the host.

Some applications of genetic engineering to plant breeding

Already there is a growing list of dicot crop species that have proved successful hosts for transformation including: alfalfa, apple, carrot, cauliflower, celery, cotton, cucumber, flax, horseradish, lettuce, potato, rapeseed, rice, rye, sugarbeet, soybean, sunflower, tomato, tobacco and walnut. In monocots, maize is leading the way, but is being followed by wheat, barley and rice.

Initial cultivar development using recombinant DNA techniques has focused on modifying or enhancing traits that relate directly to the traditional role of farming. These have included the control of insects, weeds and plant diseases. The first genetically engineered crops have now been released into large-scale agriculture (including, maize, tomato, canola, squash, potato, soybean and cotton) and other species are already 'in the pipeline'. More recently work has focused on altering end-use quality (especially oil fatty acid, starch and vitamin precursors).

Engineering herbicide tolerance into crops represents a new alternative for conferring selectivity of specific herbicides. Two general approaches have been taken in engineering herbicide tolerance:

- Altering the level and sensitivity of the target enzyme for the herbicide
- Incorporating a gene that will detoxify the herbicide

As an example of the first approach, glyphosate, the active ingredient of herbicides such as 'Roundup', acts by specifically inhibiting the enzyme 5-enolpyruvylshimate-3-phosphate synthase (EPSPS). Tolerance to glyphosate has been engineered into various crops by introducing genetic constructions for the over-production of EPSPS.

The production of plants that are resistant to insect attack has been another application of genetic

engineering with important implications for crop production. One route by which progress in engineering
insect resistance in transgenic plants has been achieved
is by using the genes of *Bacillus thuringiensis* (*B.t.*) that
produce insect toxins (so called *B.t. toxins*). *B.t.* is a
bacterium that produces a crystalline protein during
sporulation, which, when cleaved to the mature toxin
peptide, produces paralysis of the mouthparts of specific insects and so leads to their death. Thus is provides
a useful and selective means of insect control.

Transgenic tomato, tobacco and cotton containing
the B.t. gene exhibited tolerance to caterpillar pests in
laboratory testing. The level of insect damage in field trials has been similarly encouraging where tomato plants
with the B.t. gene suffered no significant damage while
non-tolerant lines were completely defoliated by insect
pests. However, an overall strategy is needed to avoid
evolution of resistance to the toxin in the insects that are
being controlled. Also recent work is investigating the
effects on, for instance, ladybirds feeding on the aphids
feeding on the transgenic plants.

Significant resistance to tobacco mosaic virus (TMV)
infection, termed 'coat protein-mediated protection'
has been achieved by expressing only the coat protein gene of virus in transgenic plants. This approach
has produced similar results in transgenic tomato and
potato, although in some other cases a similar approach
seems not to be as effective and it has been suggested
that other genes, such as viral replicase genes, might provide an effective mechanism to control virus infection
in plants.

Process of plant transformation

Before plant transformation can be used successfully in
a plant breeding programme and cultivars are developed
using recombinant DNA techniques, the following have
to be in place.

- A desirable gene must be available for insertion into
  the target host plant. Therefore a DNA clone of a gene
  that it is believed will confer a particular expression
  of the trait of interest must be developed or provided
  to the breeding programme.
- There must be a suitable mechanism to transfer the
  gene to the target plant. In dicots (and also now
  increasingly cereals and grasses) the most commonly

used vector has been *A. tumefaciens. Agrobacterium
tumefaciens* is the casual agent of crown gall disease
and produces tumorous crown galls on infected
plants. The utility of this bacterium as a gene transfer
system was first recognized when it was demonstrated
that the crown galls were actually produced as a
result of the transfer and introgression of genes from
the bacterium into the genome of the host plant
cells. (*Note*: a natural process already in existence of
introducing genetic material.)

Physical or mechanical DNA delivery systems have
been developed, and have been particularly popular
for monocots. The most common, at least initially, of
these systems involved electroporation of protoplasts
but now particle bombardment is regarded as having
wide applicability. Particle bombardment involves the
DNA to be introduced being carried through the cell
wall on the surface of a small (0.5 to 5.0 μm) piece of
metal (often gold) particles that have been accelerated
to speeds of one to several hundred metres per second.
Particle bombardment has been used for gene transfer
into a variety of target tissues including pollen cells,
apices and reproductive organs.

- A suitable construct has to be created that includes:

  A promoter region that is recognized by the host.
  These may be:

  o **Constitutive promoters** such as the 35S of
    cauliflower mosaic virus and *nos* from the nopa-
    line synthase gene which switch the gene on in all
    tissue
  o **Tissue specific promoters** such as those from
    α-amylase (specific to the aleurone), patatin
    (specific to tubers) and phaseollin (specific to
    cotyledons)
  o **Inducible promoters** such as those from alcohol
    dehydrogenase I (induced by anaerobiosis); and
    chlorophyll *a/b* binding protein (induced by light)

- A transcript terminator at the 3' end of the gene.

It might also be noted that breeders may have to re-
design the gene of interest to use codons that are the
ones more preferred by plants.

Irrespective of the delivery (vector) system used to
transform the gene into plants, foreign DNA will be
inserted into relatively few cells. A means, therefore,
must be available to select, or at least significantly

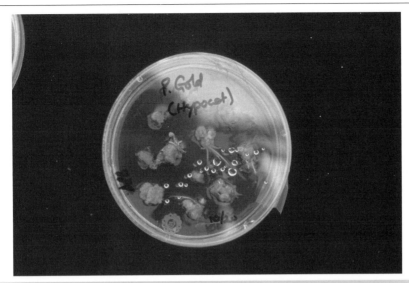

Figure 8.2   Regeneration Indian mustard (*Brassica juncea*) plantlets from callus tissues.

enrich, the cells that have been transformed. Commonly at present, this requires that the transforming gene, or a gene that accompanies it, confers some selectivity. The most usual selection agents up to now have been: **antibiotics** (mainly kanamycin or hygromycin) or **herbicides** (i.e. glufosinate or glyphosate).

Transformation allows the insertion of foreign DNA into the plant DNA of a few selectable cells. A method must, of course, be available to obtain intact mature plants from these single transformed cells. One of the biggest barriers in transformation of a number crop species is the inability to regenerate whole plants from single cells *in vitro*. In many dicots, leaf disks are transformed by infection with *A. tumefaciens*. Plantlets are then regenerated by tissue culture methods from the leaf disks. In many monocots, cultured cells, or embryos, are transformed by a suitable DNA delivery system (e.g. the particle gun) and intact plants are then regenerated from transformed cells, again, in tissue culture. Other methods that have had success are the transformation of embryogenic cell cultures or protoplasts, followed by regeneration of whole plants.

Once whole transformed plants have been produced they need to be characterized. This may be achieved by some or all of the following:

- Polymerize chain reaction (PCR) techniques can be used to detect the presence of transgene, although these techniques cannot indicate whether the gene is successfully integrated and active
- Southern blots to show the gene is present in the plant genome, and/or to estimate number of gene copies that have been inserted
- Northern blots to detect mRNA from transgenes
- Western blots or enzyme assays to show expression of the trait in plant tissues

However, this is just the start since after such tests have shown that transformed plants have the gene of interest and that the gene is functional simply means it is worth proceeding further. It then needs to be demonstrated that the gene (and expression of the trait) is stable, this generally means proving transmission of the transgene through clonal or sexual generations in order to show that any progeny will inherit and express the gene. Subsequently, the testing begins to ascertain if the expression of the gene has the desired effect on the phenotype, if any other characters are being affected directly or by the transformation process and, of course, what the actual field performance is.

Cautions and related issues

There have been a number of concerns that have arisen over the past few years as the application plant transformation technology has expanded and particularly as new transgenic crops have been released into commercial cultivation. Plant breeders need to be aware of the concerns as well as the regulations that apply to plants derived using recombinant DNA techniques as well as to other forms of gene manipulation (e.g. induced mutations). As well as the general social and environmental concerns the breeder must consider the following:

- Is the level of expression of the genetically engineered crop plant sufficiently useful to agriculture to merit the time and resources that has gone into its development, and what will need to go into any further development?
- Is there a concentration or dose effect that will further optimize the effects? Higher or lower level of expression may be produced by changing the promoter or with multiple gene insertions. If multiple copies are necessary for desirable expression then this can cause problems in the breeding programme as, effectively another polygenic trait character may have been created and will need handling in the normal way for quantitatively inherited trait.
- Will the wider use of the introduced gene lead to consequent effects in terms of the very situation it has been introduced to change. For example, all traditional plant breeders are aware of the consequent evolution in disease and pest populations of resistance to overcome single gene resistance when introduced into commercial cultivars – there is no need to repeat the same mistakes!
- Inactivation or silencing of the gene – the causes and mechanisms of which are currently being investigated. How stable will this trait be in large-scale agriculture and how will it interact (over time) with the plant's original genome?
- Care must be taken to avoid the induction of additional variation (somaclonal variation) in very generation of the transgenics. These would include obvious mutations as well as cryptic ones. (*Note*: as pointed out earlier, most plant mutations are deleterious.)
- Finally, it was at first naively believed that plant breeders would be able to take an adapted cultivar and simply transform it with a specific gene to give an 'instant' new cultivar – one that had all the previous desirable characteristics but also with the transformed trait. It is now known that this is not in fact the case. New cultivars derived through plant transformation require the same rigorous field-testing prior to release

that traditionally developed cultivars do. Multiple transformation events are necessary to ensure that one transformed plant has the desired level of expression for the altered trait, plus no deleterious epistasis interaction with the transformed gene or background changes.

MOLECULAR MARKERS IN PLANT BREEDING

Although plant breeders have practiced their art for many centuries, genetics is a subject that really only 'came of age' in the 20th century with the rediscovery of Mendel's work. Since then research in genetics has covered many aspects of the inheritance of qualitative and quantitative traits, but plant breeders usually still have little, or no, information about:

- The locations of many of these loci in the genome or on which chromosome they reside
- The number of loci involved in any trait
- The relative size of the contribution of individual alleles at each locus on the observed phenotype, except where there is an obvious major effect (e.g. height and dwarfing genes)

Theory of using markers

The idea of associating easily visualized markers in plants with loci affecting qualitative and quantitative variation in traits of interest is not new, and was first proposed by Sax in 1923. Since then a variety of contributions have been made to the general concept and theory of using mapped genetic markers for identifying, locating and manipulating genes of specific interest. The basic idea is relatively simple. If a trait or characteristic is difficult to score for whatever reason (e.g. it shows continuous variation; assessment is detailed and time consuming or the trait is only expressed after several years of growth) an easily scored marker that was determined by a locus closely associated with that affecting the character would be an attractive alternative way to monitor the locus of interest.

The concept therefore, is to use the marker locus as a point of reference for the chromosomal segment in the vicinity of the gene that is really of interest. The approach requires that alternative alleles at the marker locus match the different alleles at the locus of real interest, thus effectively marking the sections of the homologous chromosomes containing the locus that determines the particular expression of the trait we are trying to select.

The association of these marked chromosome segments with the expression of specific quantitative characters can be evaluated while allowing other chromosomal regions in the same individuals to vary at random. The aim therefore is to obtain marker genes that are closely associated with the locus determining the desirable phenotypic expression of polygenic characters such as yield or quality.

The segregating nature of F_2 populations (resulting from selfing an F_1 produced by crossing two homozygous inbred lines) often makes this generation ideal for studying quantitatively inherited characters. Investigations have also been carried out using BC_1 generations although the information obtained from this type of investigation is likely to be reduced (approximately half) of that obtainable from studies on F_2s.

With an adequate number of uniformly spaced markers (a saturated map) it is possible to identify and characterize the linkage groups, which signify the chromosomes involved. It is also theoretically possible to construct such a detailed map that the location of all major genetic factors associated with the quantitative trait might be linked rather easily and thus, by following the presence/absence of the different alleles, to describe their individual and interactive effects.

Markers in plants could assist plant breeders in the development of a better understanding of the underlying genes for characters of interest as well as providing breeders and geneticists with a powerful approach for mapping and manipulating individual loci associated with the expression of these traits. In addition, if the marker genes are tightly linked to other qualitative or quantitative characters then much of the selection in a plant breeding scheme could be carried out based on the identification of specific set of alleles at the marker loci.

The ability to identify loci which have effects on specific quantitative trait (termed quantitative trait loci – QTL) should lead not only to the ability to handle these loci in a much more deterministic manner but also to provide a more powerful means of investigating epistasis, pleiotropy and the genetic base of heterosis. So the effective use of mapped genetic markers should allow advances in cultivar development and selection procedures.

Genetic markers in plants associated with expression of morphological characters have been used for quite a long time and marker maps assembled. They have been quite well developed in a number of species (e.g. wheat, maize, peas and tomatoes) but generally had rather limited usage because of the problems of finding or generating such markers and their species-specific nature.

The characteristics of a 'good' marker system are:

- That the markers are easy, quick and inexpensive to score the phenotypes expressed
- The markers are neutral in terms of their phenotypes, and so have no deleterious effects on fitness and no effects on any other traits, including undesirable epistatic interactions with any other traits
- There is a high level of polymorphism
- They are stable in expression over environments
- Can be assessed early in the development of the plant (seedling level), and/or in tissue culture. Thus allowing evaluation without the need to grow a plant for months, or even years before it can be scored
- The scoring should be non-destructive, so that desirable individuals can be selected and grown to maturity
- Codominance in expression of the alternative alleles, so that heterozygotes can be differentiated from homozygous dominant genotypes

Types of marker systems

Any type of genetic marker that has the above properties (or many of them) may be suitable for marker-based applications in the investigation and manipulation of quantitative traits, but the question is really how closely do they conform to the ideal requirements given above.

The types of markers that can and have been used in plant breeding include:

- Morphological markers – which are basically those that you see by simply looking at a plant's phenotype, including characters such as pigmentation, dwarfism, leaf shape, absence of petals, etc. It is possible, of course, to choose ones that are easily scored but the difficulties with morphological markers are that they: cannot always be scored early in development (e.g. flower colour); are often associated with deleterious effects (e.g. albinism); are often relatively rare; their expression is not always independent of the environment in which they grow; often show dominance/recessiveness.

- Biochemical markers – such as isozyme markers. Isozymes (an abbreviation for isoenzyme) are variant forms of an enzyme, which are functionally identical, but which can be distinguished by electrophoresis – in other words when placed in an electric field. Under these circumstances the different forms of the enzyme will migrate to different points in the electric field depending on their charge, size and shape. Isozymes have been used very successfully in certain aspects of plant breeding and genetics since they: generally appear to be nearly neutral in their effects on fitness: are rarely associated with undesirable phenotypic effects on other traits: are usually free of environmental influence; and can often be extracted from tissue early in development. So they have a number of inherent properties that allows them to be used effectively for characterizing, and selection for, qualitative and quantitative characters. Unfortunately, the number of genetic markers provided by isozyme assays is not over-abundant, and they can be either co-dominant or dominant in expression. As a result, the use of isozymes as genetic markers did not allow the full potential of genetic mapping to be realized.

- Molecular markers – there are basically two systems by which molecular markers are generated and these need to be described briefly to allow an understanding of their application. The two systems can conveniently be classified as non-PCR based methods and PCR based methods. Before briefly describing each it is worth pointing out that molecular markers are simply differences in the DNA between individuals, groups, species taxa etc. Clearly the type and level of variation in DNA that we would want to examine is different depending on what level of distinction we are interested in and what questions we are answering. But the main characteristics of molecular markers are that they: are a ubiquitous form of variation; are free from environmental influence; show high levels of polymorphism; have no discernible effects on the phenotype; only pieces of tissue are required from any stage of development; and some systems show codominance.

Given the above characteristics of molecular markers, particularly their relatively unlimited numbers, it is no surprise that the advent of the possibilities of molecular markers in the 1990s was greeted with some excitement and is seen as providing a major change in the potential to exploit the ideas for using markers advocated some 70 years earlier.

Molecular markers

Non-PCR methods – DNA/DNA hybridization

The first and most widely known of these is Restriction Fragment Length Polymorphism (RFLP). Other non-PCR methods do exist, for example the use of tandemly repeated regions of DNA, known as mini-satellites or micro-satellites, but these will not be described here.

RFLP analysis involves digesting the DNA (cutting it at sites with specific sequences – there are a number of different enzymes, called restriction enzymes that cut different patterns of sequences) into fragments, which can then be separated out by gel electrophoresis (as for isozymes separating them by their differing mobilities in an electric field). To visualize their positions, they are 'blotted' onto a filter, where they are hybridized with a labelled (usually radioactive) 'probe'. The probe is a short fragment of DNA, which may be from a known gene, an expressed sequence or an unknown fragment of the genome. When the 'blotted DNA', having first denatured it to reduce it to single strands (rather the usually double-stranded state of DNA), and the probe (also denatured) are brought together, where there is an exact match in the complementary sequences they will hybridize (by hydrogen bonding) or bond. The filter is then washed to remove all the excess probe and leave only that which is now bonded with our sample DNA. If we expose the filter to X-ray film, when it is developed it will show where the probe still remains, hence where the probe has hybridized and so where there was a piece of the DNA we were investigating which had a complementary sequence. The pattern of bands obtained in this way is called the restriction fragment pattern. Using a varied combination of enzymes and probes gives a wide range of possibilities for exposing variation in the DNA sequences.

RFLP's are highly reproducible; they show codominance in their expression and are reliably specific. However, they are relatively time consuming, not easy to automate, require fairly large amounts of 'clean' DNA and, not inevitably, tend to use radioactive probes for best results.

PCR methods – arbitrarily primed techniques – multi-locus systems

The most commonly used approach is Randomly Amplified Polymorphic DNA (RAPD). The technique basically involves using a single 'arbitrary' primer in a PCR reaction. The primer is basically just a short stretch of DNA. The basic ingredient of the PCR reaction is DNA polymerase, an enzyme that enables the copying of a duplicate molecule of DNA from a DNA template, and is commonly *Taq* polymerase a thermally stable DNA polymerase. The primer anneals to the complementary sequences in the DNA we are investigating and 'primes' the polymerase amplification. So the events which occur are:

- Isolate the DNA from the organism of interest
- Put in thermal cycler with the primer and polymerase
- Denature the double-stranded DNA by heating
- Anneal primers to initiate extension of sites flanking region by cooling
- Primer extension – synthesis of DNA strands complementary to the region between the flanking primers with *Taq* polymerase
- Repeat the three cycles, above, basically doubling the specific region determined by the primer on each cycle – so quickly enriching the mixture to be almost purely pieces of this one region of DNA – the basis of PCR
- The products are separated on agarose gel, commonly, in the presence of ethidium bromide and visualized under ultraviolet light

The advantages of RAPDs are that it requires only small amounts of DNA, it requires modest equipment (thermal cycler and electrophoresis equipment); and no prior knowledge of the gene or DNA sequence is required. It is fast and relatively inexpensive. However, the results can be variable depending on slight changes of the PCR conditions or ingredients and the markers show dominance

More reliable methods that have been developed are: Amplified Fragment Length Polymorphism (AFLP) and this is not only more repeatable but also gives much higher frequency of bands; and inter-simple sequence repeats (ISSR or anchored micro-satellites). However, the details of these are beyond our present remit.

PCR methods – site targeted techniques – single locus systems

Rather than using arbitrary primers, it is possible to specifically design primers to be used in PCR. There are a number of possibilities to design primers but one such approach is Sequence Tagged Micro-satellites (STMS). Micro-satellites are simple sequence repeats which are found around the genome and are generally quite variable in exact base pair composition. If one pictures these at different places in the genome, the DNA 'flanking' these regions will be different depending on where they are (i.e. the site at which they are found will be unique). So you can 'fish' for these with simple repeats, then sequence the bands and design primers with the main part being simple repeats but the ends being other unique 'tags'. This allows the production of much more robust markers to be generated but with the advantage of the PCR technology.

Uses of molecular markers

Molecular markers can therefore be used to:

- Identify cultivars (DNA finger printing), to differentiate one cultivar from another (perhaps one already released), or to be able to prove proprietary ownership of specific cultivars. If you have a modest set of markers it is possible to produce a 'DNA finger-print' which is unique (or nearly so) and so be potentially used to identify that particular genotype. Similarly using the same principal it is possible to identify DNA that is not supposed to be there and so can be used to ensure that a particular cultivar is pure and free from contaminants. A further possibility is afford by the potential to assess how diverse genotypes are at the DNA level and hence assess their level of difference (genetic distance) if used as parents (so e.g. parents of hybrid cultivars).

- Marker assisted backcrossing. When a gene of interest can be shown to be linked to a molecular marker, then assessment of the marker can help accelerate the backcrossing process. Mature plants would not need to be grown to identify which backcross individuals carry the allele of interest. This is particularly helpful where, although determined by a major gene, the phenotypes are difficult or time consuming to detect or are expressed later in development (e.g. fruit colour). Molecular markers can identify which of the back cross progeny have better restoration of the rest, or background, of the genome of the recurrent parent.

- They can provide breeders with vital information about the legitimacy of any cross but particularly if a supposed wide cross (or interspecific cross) is genuine or the result of an unfortunate illegitimate pollination. Indeed when used in conjunction with cytogenetics information can give very precise information about what chromosome or parts of chromosomes are present in interspecific hybrids or generations derived from such hybrids.

- When a number of markers have been generated then they can be used to build a map of the genome and hence provide much clearer ideas of the positions on chromosomes of different genes and so determine the associations that might be expected between simply inherited traits. Thus helping to determine the selection strategy that will be most applicable.

- QTL can be followed by the behaviour of an associated marker locus. The more detailed a linkage map exists the simpler it is to associate quantitative traits with existing markers. Single QTL markers, should be linked at 2 cM to be effective and to have two flanking markers is even better. The probability of a double cross over between the markers then being remote. An initial map means that markers can be selected which roughly cover the genome (perhaps at least 2 markers per chromosome arm) to look for basic co-segregation. Once a chromosome segment has been identified this area can be concentrated on in terms of using all the markers that are available in this particular region. However, the difficulty remains in assessing the quantitative trait expressions accurately and in ways that are relevant to the agronomic circumstances that the cultivars will finally be grown in. Genotype × environment interactions could pose as large a problem in QTLs as it does

in traditional evaluation and selection. QTLs, however, might offer plant breeders an opportunity to obtain a better understanding of the genetic basis of genotype × environment interactions, epistasis and heterosis. Also, it is clear that amongst the quantitative variation exhibited for many traits there are some regions of DNA, which determine rather large parts of the variation that we observe – if these could be handled effectively the effort that was saved could be focused on the non-defined regions.

Problems with markers

In many instances using molecular marker techniques (say for selection) is basically more expensive and more technically demanding than other selection options. It is therefore not really cost-effective to set the necessary laboratory facilities and trained staff to handle a few crosses or perhaps a situation where the profit returns on the breeding are low.

Finding a molecular marker, which is associated with a major gene of interest or a QTL, is not always too difficult but ensuring that it is close enough not to be lost by subsequent recombination is more difficult. Also the applicability of the marker combination over a range of crosses rather than just a specific one is also a concern that takes time and effort to ensure.

Nevertheless the exploitation of QTLs offers great potential that has yet to be realized in practical terms. Developing good and reliable QTLs will require a great deal of well designed and accurate field-testing. As already noted, genotype × environment interaction may pose as large a problem in QTLs as they do for traditional selection. Finally, there needs to be even better repeatability between results obtained by different research teams. Different researchers sometimes identify different loci to be responsible, in QTL analyses, for the major differences in expression of the phenotypes for the character of interest. Some of these differences will reasonably be ascribable to the fact that different alleles are segregating in different crosses or being expressed at different levels in different circumstances – note the similar problems with heritability estimates! But there are also technical differences, which need to be corrected before the true potential of QTLs can be realized.

THINK QUESTIONS

(1) Outline the potential impact, advantages and disadvantages of the following techniques in plant breeding: mutagenesis haploidy; molecular markers; genetic engineering (transformation); and interspecific hybridization.

(2) Name the four types of mutation that can occur in plants and briefly describe the features of each type. Outline two difficulties that might be problematical when using mutagenesis in a plant breeding programme.

(3) Interspecific and intergeneric hybridization can sometimes be useful technique in plant breeding by introgression of characters and genes from different species. In an interspecific cross between *Brassica napus* and *B. rapa*, there was no evidence that any *B. napus* egg cells had been fertilized. List three reasons that could have caused this non-fertilization.

(4) Haploidy is used in many pure-line breeding programmes. Outline four reasons why using haploidy techniques as a routine procedure in a practical barley breeding programme may not be feasible.

(5) Outline any differences between morphological and molecular markers as used in plant breeding. List four applications of markers in a plant breeding programme.

(6) List four vectors that can be used to transform plants with genes from other organisms and briefly describe how each vector system works. Briefly describe two problems that plant breeders may encounter in developing transgenic cultivars.

9

Some Practical Considerations

INTRODUCTION

As noted at the beginning of this book, plant breeding demands a range of skills including good management and a multitude of other scientific disciplines in combination, to achieve success. Plant breeding operations and evaluation of plant breeding lines will be conducted in laboratories, glasshouses and field situations. This final section attempts to outline some of the practical difficulties in a plant breeding programme. Sections covered in this chapter examine: experimental design, including the types of designs suitable for different parts of a plant breeding programme; glasshouse management and field management and the applications that can be covered and managed using computers. Finally, this chapter considers some of the practical considerations of the actual cultivar release procedure.

EXPERIMENTAL DESIGN

It has been stressed previously that the basic operations of cultivar development can, for simplicity, be divided into three stages: producing genetic variation, selection among recombinants for desirable new cultivars with specific characteristics, followed by stabilization and multiplication. The following few sections are concerned particularly with the middle one of the three processes.

The aim in selection is to identify recombinants, which are genetically superior to existing cultivars. Superiority can be achieved by increased productivity (e.g. increased yield or better end-use quality), by making productivity less variable (e.g. reduced risk of crop failure by introduction of disease, insect or stress

tolerance or resistance) or increased profit (e.g. reducing input costs by incorporation of disease resistance).

In each of the cases it will always be necessary to evaluate the performance of breeding lines for qualitative and quantitative characters. In some instances it is possible to select and screen for single gene characters without the complication of interaction with environmental factors. However, it is accepted that virtually all quantitatively inherited characters (most often the ones with greatest commercial value, e.g. yield, quality and many durable disease resistances) are highly modifiable by the environment. Consider the observation of a single plant; the aim is to minimize the non-genetic effect from the equation:

$$P = G + E + (G \times E) + \sigma_e^2$$

where P is the phenotypic expression, G is the genotypic effect, E is the effect of environmental variables, $G \times E$ is the effects attributable to the interaction of the genotype with the environment effects and σ_e^2 is a random error term associated with a single observation. In the evaluation of breeding material it is only possible to observe the phenotype (a combination of genotypic and environmental effects). The aim is to determine the genetic potential of each breeding line, and hence it is necessary to either estimate or minimize the environmental and error effects. To achieve this demands careful use of a number of **experimental designs**.

Running a plant breeding programme is no different from organizing a whole series of scientific experiments and therefore all aspects of the operation should be treated with the same care and detail that individual experiments should be planned and handled. Good experimental design leads to knowledge of the accuracy of the data on which evaluation and selection are based. The quality of information collected in a plant breeding programme is the key factor in determining the success of the scheme, whether it is one based solely on traditional techniques or it incorporates molecular based technology.

It is common to evaluate breeding lines (**test entries**) in comparison with existing cultivars (**controls** or **checks**) within the same trial. In some cases a single cultivar is used, but more often several cultivars are included in the evaluation trials. The choice and number of control entries is largely dependant on the range and number of cultivars that are presently grown in the target region for the new cultivars, the type of trial and the number of evaluations that are to be made. For example, an evaluation trial may contain the highest yielding cultivar available to compare yielding performance, the best quality cultivar to provide a baseline for quality, a cultivar with disease resistance to evaluate response under disease pressure, and so on. It is always desirable to include as many control entries as possible within reason for the extra effort involved. It should be noted, however, that evaluation trials can be costly and that the cost of an evaluation trial is often directly related to the number of total entries that are included. If several thousand breeding lines are to be evaluated then it may be unwise to include only a few control entries. If, however, only a few lines were to be considered then it would be unwise to include many hundreds of control plots. A simple rule of thumb which is often useful is that the number of control plots (not always entries) should be about 1/10th of the total number of trial entries if between 1 and 200 breeding lines are to be tested, and up to 1/20th of the total number in the trial if more than 200 lines are under evaluation.

A wide spectrum of possible designs is available but only a limited number will be detailed here, namely:

- Unreplicated designs
- Randomized complete block designs
- Factorial designs
- Split-plot designs

Unreplicated designs

Unreplicated designs, as their name suggests, are experimental designs where test entries are not replicated and so appear only once. There are, however, several (or indeed many) different options even when single replicate designs are used. These include:

- Non-randomized designs without control entries, where genotypes that are to be evaluated are arranged in plots in a systematic order (e.g. numerical order, alphabetical order etc.) (Figure 9.1 (a)). Only test genotypes are evaluated and there are no control (check) cultivars grown at the same time. It is not possible to obtain any estimate of error from this type of design or make any direct comparison with known cultivars.
- Non-randomized designs with control entries, where the test entries are arranged in plots in a systematic order (as above), but control cultivars are inter-spaced amongst the test entry plots (Figure 9.1 (b)). The control cultivars can be arranged in a systematic order (e.g. every 20 plots), or they can be allocated to plot positions at random. In most cases several control cultivars are included. Each control entry may also be replicated more than once in the whole design. Multiple entry plots of control cultivars can often be useful to determine an estimated error variance for the overall field trial.
- Randomized designs without control entries, where the test entries are arranged within the trial at random but no control entries are included (Figure 9.2 (a)).

(a) No randomization, without controls

49	50	51	52	53	54	55	56
41	42	43	44	45	46	47	48
33	34	35	36	37	38	38	40
25	26	27	28	29	30	31	32
17	18	19	20	21	22	23	24
9	10	11	12	13	14	15	16
1	2	3	4	5	6	7	8

(b) No randomization, with controls

37	38	C.1	39	40	41	C.2	42
31	C.3	32	33	34	C.4	35	36
C.1	25	26	27	C.2	28	29	30
19	20	21	C.3	22	23	24	C.4
13	14	C.1	15	16	17	C.2	18
7	C.3	8	9	10	C.4	11	12
C.1	1	2	3	C.2	4	5	6

Figure 9.1 Non-randomized single replicate plot designs without control entries (top) and with control entries arranged systematically throughout (bottom).

(a)   Randomization, no controls

5	22	50	47	34	36	52	24
31	40	18	38	28	3	49	25
43	32	7	41	10	37	19	1
14	56	29	55	12	54	35	30
26	9	45	48	27	44	2	42
53	20	51	17	6	23	21	13
11	39	46	15	15	4	33	8

(b)   Randomization, systematic controls

38	13	C.1	41	29	5	C.2	32
33	C.3	1	24	4	C.4	18	30
C.1	21	39	23	C.2	36	26	9
40	8	2	C.3	15	16	27	C.4
42	12	C.1	35	3	19	C.2	11
20	C.3	34	28	37	C.4	25	31
C.1	7	22	14	C.2	6	17	10

Figure 9.2   Randomized single replicate plot designs without control entries (top) and with control entries arranged systematically throughout (bottom).

- Randomized designs with control entries, where the test entries are randomly allocated plot positions within the trial (Figure 9.2 (b)). Control entries can also be randomized throughout the design (and often replicated in more than one plot) or they can be arranged in a systematic order (e.g. every 5th plot) with again the option of having replication only for the control entries.

The efficiency of evaluation trials will always be increased by randomization, and non-randomized trials should be avoided if at all possible. Similarly, it would be very unwise to organize any breeding evaluation trials without including any control entries against which the test lines will be compared. Without these considerations the trials are generally uninformative and often misleading.

In the early generations of a plant breeding scheme, there may be many hundreds or thousands of genotypes to be tested, each with only a limited amount of planting material. In many breeding programmes, the first 'actual' field trials are conducted on head-row plots, where each plot has resulted from a single plant selection the previous year. Where thousands of lines are to be tested, it may be extremely difficult to completely randomize each individual head-row, but randomization at this early generation stage can greatly increase efficiency. One option is to utilize nested designs. For example, say that a canola breeding programme has 200 cross combinations to evaluate and that there are 100 individual single plant selections taken at the F_3 stage. Therefore there would be 2000 F_4 head-row plots that would be planted in the field. A randomized complete block (with control entries) would be very large. In addition, from a practical aspect, it is often difficult to examine a single row plot, on-its-own. As an alternative the 200 crosses could be randomized into five replicate blocks, and the 100 single plant selections are grown as rows within cross blocks. Each cross, therefore, would be represented by five sub-blocks (groups) of 20 head-row plots (grown adjacent), and replicated five times throughout the whole trial.

If control entries are arranged in a systematic order it will be possible to make direct comparisons of individual test entries to the nearest control plot, which can have advantages. For example, it makes possible the analysis of the data collected using **nearest neighbour** techniques, where plot values are adjusted according to the performance of appropriate surrounding test entries.

Randomized designs

It is possible to obtain an estimate of error variance from single replicate designs which have multiple entries of chosen control cultivars. However, it is more common, if possible, to replicate both test lines and control cultivars in order to have a better estimate of the average performance of each entry, along with the variance in its performance, and also to obtain a better and more representative overall estimate of error variance.

Completely randomized designs

If there is no knowledge of fertility gradients or other environmental variation, which exists within a test area, many suggest that complete randomization be used to identify superior breeding lines. In such a design each of the test and control entries are allocated at random to specific plot positions (Figure 9.3 (a)). Each entry is repeated a number of times according to the required number of **replicates**. The error variance is estimated from the variance between replicate test entries.

(a) Four Replicates ~ No Blocking

16	16	14	9	13	16	15	3
1	13	6	11	5	12	10	4
10	12	2	7	6	15	11	2
2	5	13	7	1	7	6	12
4	14	3	13	8	6	9	1
11	4	3	8	7	14	15	10
8	12	14	4	2	5	16	8
3	9	15	8	10	11	5	1

(b) Four Replicates ~ Blocking

9	14	5	13	9	16	11	5
11	7	16	1	3	13	8	6
15	2	3	12	2	15	14	4
10	6	4	8	1	10	7	12
6	13	10	16	14	6	9	8
14	3	1	8	12	11	15	4
12	15	9	5	1	16	3	15
7	2	11	4	13	2	10	5

Figure 9.3 Completely randomized block design (top) and randomized complete block design (bottom).

Factorial design

L.2 T.3	L.3 T.1	L.1 T.3	L.4 T.3	L.2 T.3	L.4 T.1	L.1 T.1	L.4 T.3
L.3 T.3	L.1 T.1	L.2 T.1	L.2 T.2	L.3 T.2	L.2 T.2	L.4 T.1	L.1 T.2
L.3 T.2	L.4 T.1	L.1 T.2	L.4 T.2	L.2 T.1	L.3 T.3	L.1 T.3	L.3 T.1

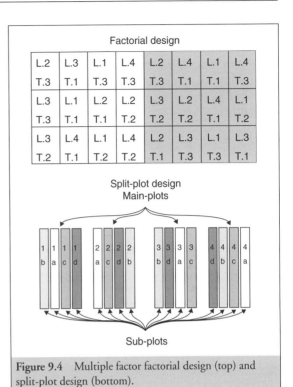

Figure 9.4 Multiple factor factorial design (top) and split-plot design (bottom).

Randomized complete block designs

Although there are often merits in choosing a completely randomized design, a more common design (probably the most common design) used by plant breeders is a randomized complete block. In these designs, the total area of the field tests is divided into units according to the number of required replicates. Each unit is called a block. Each of the test and control entries are randomly assigned plot positions within each block (Figure 9.3 (b)). In the cases where there are distinct fertility gradients or other differences between blocks, then these can be estimated and subtracted from the error variance. It is possible therefore, to obtain a more accurate estimate of the error variance. Blocking does not necessarily need to be different areas within a field trial. Different blocks in a randomized complete block design could, for example, be different days of testing (where it is not possible to test all replicates in a single day).

Factorial designs

Single replicate designs are often referred to as single dimension designs and randomized designs are called two-dimension designs. In many cases it is important to simultaneously evaluate a number of breeding lines with regard to their response to different treatments. These types of experimental designs are called multidimensional designs or **factorial designs**. To illustrate factorial designs consider the example where there are only four breeding lines to be tested (L.1 ··· L.4) and the performance of each is to be evaluated under three different treatments, or factors (T.1 ··· T.3). Each genotype entry is grown with each of the different treatments. Overall, there are therefore $4 \times 3 = 12$ entries. These are arranged at random as illustrated in Figure 9.4 (a). In the example only two replicates are illustrated. In practice more than two plots of each test unit would be grown to ensure the necessary level of replication. Replicated factorial designs can be completely randomized or each replicate can be blocked.

Analysis of factorial designs allows estimates of differences between test entries and between treatments compared to an estimated error. These designs also allow evaluation of any **interaction**, which may exist between test entries and treatments. To illustrate consider the performance of two test lines (A and B) each evaluated under low and high nitrogen conditions. If the yield performance of the two lines follows the pattern where entry A is higher yielding than entry B at both nitrogen levels, it is said that there is no interaction. If, however, entry A is highest yielding at high nitrogen levels but entry B is highest yielding at low nitrogen levels then there is said to be interaction between genotypes and nitrogen levels. The significance of interaction is tested from analysis of variance. It should be noted that in testing genotype × treatment interactions if there are no changes in genotype ranking then although formally an interaction will be detected the implication for plant breeding is minimal, unless other treatments (maybe outside the range tested) are envisaged as being likely.

Split-plot designs

In some cases a breeder is interested in estimating the difference between the effect of one factor and the interaction of that factor with a second factor while having lesser interest in variation within the second factor in its own right. In plant breeding for example, it is well established that higher nitrogen (within limits) applied to cereal crops will result in higher yield. Genotypes × nitrogen studies are routinely carried out, not to determine if there is an average yield increase with increased nitrogen (as this has already been established). The primary goal is to determine the differences between genotype variation and the interaction between genotypes and nitrogen levels. In these cases a special type of factorial design called a **split-plot design** is commonly used.

Split-plot designs divide the total area of a test into **main blocks**, **sub-blocks**, **sub-sub-blocks**, etc. First, the main blocks are arranged at random, and then factors within sub-blocks are arranged at random within the main blocks, factors of sub-sub-blocks are arranged at random within sub-blocks etc. A simple split-plot design is illustrated in Figure 9.4 (b). The numbers 1, 2, 3 and 4 are main blocks and the letters a, b, c and d are sub-blocks. Only a single replicate is shown in the figure, but as before other replicates would be similar and would always be blocked.

The analysis of variance produces two errors for use in F tests. The error #1 is estimated by the sub-blocks within main blocks × replicates effect while the interaction effects would be tested against the between main blocks × replicates interaction.

GREENHOUSE MANAGEMENT

A large proportion of the tasks that are necessary in the early parts of a plant breeding programme can often be carried out in a greenhouse. An integrated greenhouse system is not essential for a successful varietal development programme. However, many of the operations can be carried out more effectively if a greenhouse system is conveniently available.

Greenhouses come in many different shapes and sizes and can be constructed from many different materials including wood, aluminium, glass, plastic, polythene. The actual design of these systems will not, however, be covered here. A greenhouse can simply be considered as a relatively large area where there is some control of environmental conditions such as soil type, irrigation management, nutrient management, lighting and temperature.

The operations, which can be carried out in a greenhouse with regard to a plant breeding programme, include:

- Artificial hybridization
- Seed increases of breeding lines, including progressing to homozygosity
- Evaluation of characters, which are difficult to control under field conditions

Artificial hybridization

It is possible to carry out artificial hybridization under field conditions, however, many breeding schemes use greenhouse facilities for this task because it is easier to control the conditions necessary to ensure cross-pollination between chosen parents (Figure 9.5). Also, it can often be possible to achieve cross-pollination out of season in greenhouses and usually it is easier to prevent unwanted illegitimate cross-pollination.

The method used for cross-pollination will be dependent on the crop species involved, whether the crop is out-crossing or self-pollinating. The major goal in artificial hybridization is to ensure that the seed produced is in fact from the particular, desired paired parent combination. Therefore steps must be taken to make sure that seed has not resulted from an unwanted self-pollination or from an accidental cross-pollination that is not intended by the breeder.

Figure 9.5 Artificial hybridization in canola breeding (a) and wheat breeding (b). Note that pollination bags cover racemes and ears that have been pollinated so as to avoid unwanted crosses.

Artificial pollination therefore demands that naturally inbreeding lines (or lines which are self-compatible) be **emasculated** to avoid self-pollination. Emasculation in most crop species can be achieved by manually removing the male plant parts (i.e. anthers) before they are mature and pollen is dehisced. In some cases it is possible to use chemical emasculation where specific chemicals applied at the critical growth stage will render the plants male sterile. Chemical emasculation is, however, not widely used in routine breeding and mechanical emasculation is used almost exclusively employed as a means of avoiding selfing in crossing designs.

After the chosen female plants have been emasculated, within a few days pollen from the male parents can usually be applied to the receptive female stigma. Pollen can be transferred manually, often using a small paint brush or by removing dehisced male parent anthers and brushing pollen onto the female stigma. Cross-pollination can also be achieved simply by having emasculated females grown in close proximity to male flowers and allowing pollen to naturally pass from male to female. When this is to be done it is common to place emasculated female flowers and pollen fertile male flowers together within a pollination bag to ensure that the designed hybridization occurs and to avoid the female from being pollinated by stray pollen which may, for example, be blown in the air. If necessary, within these bags, suitable pollinating insects can also be placed to help pollination efficiency

In several crop species there are self-incompatibility systems, which have developed naturally to maximize heterozygosity of plants within the species (e.g. as exists in many *Brassica* species). Similarly, many crop species have male sterility systems (either nuclear or cytoplasmic in inheritance), which can be utilized in cross-pollination systems. In both these cases it is not necessary for the female parents to be emasculated to guarantee cross pollination.

Irrespective of the breeding system, it is common to place pollination bags over flowers either prior to pollination (to avoid unwanted crosses) or after pollination (to ensure that no further pollination takes place) (Figure 9.5). It should be remembered that bagging crosses will be time consuming and that if not done carefully can have an adverse effect on the potential success of the artificial hybridization. Damage can be caused during the bagging operation or the bag may create an environment unsuitable for seed production. It also is important to label carefully at each stage, otherwise the origin of any seed produced may be in doubt.

With many crop species, particularly crops which are clonally propagated and where the end product does not involve the botanical seed (e.g. potato, banana, sugarcane) it is not always easy to have parental lines develop sexually reproductive parts. In a number of instances flower induction can be achieved by manipulation of environmental conditions by adding or reducing nutrient levels manipulation of day length or by artificially controlling the natural source–sink relationship.

For example, in potato, many past breeders have specifically selected breeding lines, which rarely produce flowers with the idea that energy put into sexual reproduction would detract from tuber yield. Flowering can be induced in some genotypes by planting tubers under long day conditions and having plants develop to maturity in shorter days. Enhanced flowering in potato can also be achieved by 'growing on a brick' where parent tubers are planted on building bricks and covered with soil. At the stage when tuber initiation occurs the soil is washed from the mother tuber and newly initiated tubers are removed, hence offering greater resources for flower development. A similar effect can be achieved by grafting potato shoots onto tomato seedlings. Applying high levels of nitrogen at particular growth stages can sometimes increase the duration of the flowering period.

In other crops (and sometimes also in potato) reduced levels of nutrients cause stress to parental plants which can induce flowering, which would otherwise not occur under optimum conditions.

Finally, irrespective of crop or breeding system it is always desirable to have multiple and sequential plantings of parents that are to be used in crossing designs. Genetically different parents will of course flower and dehisce pollen at different times and multiple plantings will increase the possibility of achieving all hybrid combinations planned.

Seed and generation increases

If hybridization is carried out between two homozygous parents then the F_1 plants will be heterozygous at all the loci by which those parent lines differ and all plants will be genetically identical. It is therefore common practice to go from the F_1 populations to F_2 under glasshouse conditions. This tends to maximize the use of F_1 seed because of the high levels of germination and survival that can be achieved. If F_1 populations are grown under field conditions it generally requires greater quantities of hybrid seed. This is disadvantageous since the cost of producing F_1 seed is usually high, because it involves emasculation followed by hand-pollination, in relation to simply bagging the F_1 to allow selfing to produce the F_2.

With many annual (and some biennial) crops it is possible to grow more than a single generation each year, therefore greenhouses can be used to reduce generation times and hence increase the speed to homozygosity. Single seed descent used in spring barley, where plants are grown at high density and with low nutrition, can be used to increase F_1 populations to F_3 populations within a single year (i.e. three generations in 12 months).

At the advanced stages of a plant breeding scheme, greenhouse growth can be utilized to increase advanced selections under controlled conditions prior to producing breeders' seed. This can be particularly useful in crops which are grown as true breeding, inbred lines but in which a relatively high frequency of natural out-crossing occurs (e.g. *Brassica napus*).

Tissue culture techniques are becoming a routine part of many plant breeding schemes. Plants rarely can be transferred directly from *in vitro* growth to field conditions without involving an intermediate greenhouse stage. Here the greenhouse stage could involve an intermediate operation where plants are weaned from *in vitro* to *in vivo* sterile soil mix, allowed to develop and are later transplanted to the field. Alternatively the greenhouse can be used to produce seed (or tubers) from plants that have previously been grown *in vitro*.

Evaluation of breeding lines

One advantage of growing plants under greenhouse conditions, rather than field conditions, is related to environmental control. Control of the environment can be critical to guarantee epidemics of pests or disease or to evaluate stress factors, to allow resistance screening. There have been a number of studies that have resulted in protocols suitable for evaluating plants under glasshouse conditions.

Disease and pest testing involves subjecting segregating breeding populations to a disease or insect and selecting those plants, which show resistance. Examples include spraying barley seedlings with a suspension of mildew spores and screening for resistant lines, spraying potato seedlings with a spore suspension of late blight or early blight and recovering the seedlings that are not killed. These tests are often more effective if there is good environmental control, such as is provided in a greenhouse. This helps to guarantee that the results are repeatable and the particular pathogens are allowed to increase and indeed infect the plants. It also allows control of the disease when it is time to stop further infection

Screening breeding lines for abiotic stresses can also be achieved under greenhouse conditions if the environment can be controlled in a repeatable and relevant manner. Stress screening has been shown to be reliable to such factors as tolerance to nutrient deficiency, drought, salinity and heat where it is not always possible or easy to control the relevant environmental factors involved under natural conditions in the field.

It should, however, be noted that evaluations designed to be carried out under greenhouse conditions must first be compared to results that would have been achieved under natural field condition. There have been numerous cases where selection has been carried out under controlled conditions and later found to bear little, if any, relationship to what subsequently is experienced under field conditions.

Environmental control

Artificial lighting (fluorescent and/or incandescent) is nearly always necessary to achieve maximum use of greenhouse space. Lighting is, however, expensive both to install and maintain, particularly if different day lighting regimes are required. When, however, lighting is available, it usually allows the greenhouse to be utilized throughout the whole year.

If plants are to be propagated in the greenhouse throughout the year it will also be necessary to have a suitable heating and/or cooling system. A range of different types of systems is available and these cannot be adequately covered here. However, it should be noted that all the types require a relatively high cost to install and operate. Thus it is usual to expect to have to justify the costs in terms of likely returns of, for instance, increased numbers of generations, effectives of tests etc. Good control of temperature is of course important if healthy plants are to be propagated. A particular example in which temperature control is often needed is in biennial crops where plants require vernalization (chill treatment) before they will flower. Plants grown under greenhouse conditions can be vernalized outside the greenhouse (e.g. in a growth chamber or cold room) but this will involve moving plants between

facilities which can be time consuming and expensive if the number of plants involved is large.

Growth within greenhouses requires artificial irrigation. Irrigation can be by hand, which allows for some flexibility but does not usually allow for complex irrigation management systems. Automatic irrigation is usually preferable and can be of three forms:

- Above plant irrigation (or misting) where plants are sprinkle or mist irrigated from above. This can be relatively inexpensive but can cause problems if plants are tall. Above plant irrigation can also increase the risk of plants becoming infected by fungal diseases where leaf moisture is necessary for infection to take place. It can also be a problem in generating leaf scorch in strong sunlight.

- Below plant irrigation where plants are irrigated by capillary action by having moist or wet material below the plant pots. Below plant irrigation avoids the above-ground plant parts becoming wet although it can be difficult to establish young plants and maintain very large plants with such a system alone and it is sometimes necessary to hand water as a supplement to the system.

- Drip irrigation where each plant pot is individually irrigated directly into the soil by a drip line. There are several different forms of drip irrigation and this system offers greatest flexibility over all others. This system is, however, the most expensive to install and is not always available in all plant breeding greenhouses. Since the system requires that individual drip lines are located in each plant pot, there can be some restriction on the number of plant units that can be grown, so this need to be carefully considered when setting the system up.

All methods of irrigation offer the possibility of applying nutrients along with the water and so they can be provided 'continuously' thus enabling more optimized growth of plants over other methods of nutrient/fertilizer application.

Disease control

Unless disease is to be deliberately encouraged, as in the case of a screening scheme, it is desirable to avoid as many diseases and pests as possible in a breeding greenhouse. The best results are invariably achieved when the plants are as healthy and disease-free as possible. Crop failure in a greenhouse as a result of plant pests (mainly insects) or disease can carry a high cost and should, of course, be avoided if at all possible.

Disease and pest control can be achieved by adopting good management practices, including sensible breaks in production along with appropriate sterilization strategies. However, the application of chemical insecticides and fungicides is also a frequently needed practice. Application can be by spraying plants or pests or by fumigating a whole area within the greenhouse. The main advantage of chemical control of disease and pests are that they can be applied in anticipation that a problem will exist. Therefore they offer **preventative** disease and pest control. The disadvantage is that many of these chemicals are indeed harmful both to humans and other plant and insect life and it is therefore always desirable to minimize their use.

There are now many types of biological controls that can be used to control insect pests within a greenhouse. A well-known example is the release of ladybugs (ladybirds) which are natural predators of aphids into greenhouses. There are many other predator insects available that can offer effective control of other insect pests. A sample of specific predator types available and the pest they attack include: *Amblysieus cucumeris* against thrips; *Aphidoletes aphidimyza* against aphids and *Encarsia formosa* whitefly, while ladybugs and green lacewings are used as general insect predators.

The major difficulty of biological predatory control relates to the fact that the pest must in fact be there, even if at a low level, before the predators are released (otherwise how will they survive!). It is therefore difficult to avoid some insect damage and almost impossible to achieve complete preventative control.

The risk of soil-borne diseases can be avoided (or at least substantially reduced) by using only sterile soil, or soil mixes, in the greenhouse. However, unless an inert, synthetic soil substitute is used (e.g. 'Perlite' or sand/gravel/Perlite) the possibility that disease will occur as a result of infected soil cannot be entirely avoided. Often the sterilization procedure fails to remove all disease or fails to kill weed seeds. In addition if peat moss is used in soil mixes it is almost impossible to ensure the mix is free from insect pests that have a reproductive cycle in the peat moss.

Achieving good disease and pest control in greenhouses can be achieved by other means. For example,

good insect-proofing throughout the house will reduce the risk that insects will enter the greenhouse. However, it should be borne in mind that people are very effective spreaders of plant disease in greenhouses. Personnel from the breeding programme are likely to be in contact with plants outside the greenhouse (i.e. will visit field plots) and so there is a great risk that these staff will transmit disease or carry in insect pests prevalent to the crop with them while visiting the glasshouse. Simple rules, such as any greenhouse operations are carried out first thing each day and other field tasks are done later, can help in reducing disease incidence and spread.

Plant viruses can cause particularly serious problems in plant breeding schemes as many virus diseases are transmitted through the planting material (e.g. seed viruses in cereals and tuber borne viruses in clonal crops). Many viruses can be eliminated by avoiding the virus vectors, which are often insects (particularly aphids). Workers in the breeding programme can also be responsible for carrying insect vectors into greenhouses on their hands or clothes. Again the risks of infection can be reduced by applying simple rules (e.g. protective clothing, sterile gloves etc.).

Economics

Despite the attraction of greenhouses as an integral part of any plant breeding programme, there is no doubt that this facility can be responsible for a high proportion of the overall cost of operating a breeding system. In addition, due to the high cost of building and maintaining greenhouse facilities the actual space available will be limited. In the practical world (the one in which we unfortunately all live) economic use of greenhouse space will become a major factor.

Plants in greenhouses are grown either in pots (or some other individual unit) or in beds (where many plants are propagated together). The size of pot used (or the plant density in seedling beds) will have a large influence on the number of plants that can be grown in a unit area. It is therefore necessary to choose a density pot size that will allow good plant health and growth. If small pots are used then more plants can be propagated at lower cost. If they are, however, too small, then plant health or reproductive efficiency can be affected.

It is necessary to allow access to plants grown in greenhouses. Increased efficiency of greenhouse space can be achieved by **rolling benches** where plants are grown on benches that can be easily moved to allow access, but minimizes the greenhouse space that is allocated to walkways. Rolling benches can, however, cause problems in cases where plants are tall and require staking and tying or else they will fall over and be damaged. In addition, rolling benches can increase the need for uniform lighting over the whole greenhouse area rather than only over designated growth areas or static benches.

Experimental design in the glasshouse

One final note on the use of greenhouses and plant breeding relates to experimental design. Many believe that the conditions in greenhouses are such that there is uniformity in soil type, lighting, irrigation etc. In comparison to conditions that may prevail in the field, there may indeed be less environmental variation in a greenhouse. Despite this, it should be noted that there will be differences nonetheless between, say, plants next to the glass and those in the centre of the house. Therefore **all** experiments grown in greenhouses should be treated with a clear understanding of the fact of variability in environmental variables exists, and therefore good experimental design, replication and randomization will be as important in greenhouse experiments as in other situations.

FIELD PLOT TECHNIQUES

A large proportion of the work in a plant breeding programme is carried out using field trials. The aim of plant breeding is to develop superior cultivars that are genetically more adapted than the cultivars that are already available. New, and old, cultivars are grown within agricultural systems on a large scale. For example, wheat grown in the Pacific Northwest is grown in fields, which cover many hundreds of acres. Obviously it is not possible to evaluate the many thousands of potential new cultivars in a plant breeding scheme on the large field areas that they will eventually be grown if successful. The aim therefore of field trialling is to **predict** how each genotype would perform **if they were grown on a large acreage basis**.

In order to grow accurate and representative field plot trials it is necessary to first determine the way that the crop is grown in agriculture and to try and use this as a basis for the practices used in the plot trials. Factors that need to be determined include:

- Land preparations
- Seeding rate, final plant density and depth of planting
- Nutrient levels and when nutrients are available (pre-plant and/or post emergence)
- Irrigation management
- Timing of operations such as planting and harvest windows
- Chemicals available, for example what insecticides, fungicides or herbicides are registered for use on the crop, at what rates they are applied, what seed treatments are?
- Regions where the new cultivars will be targeted

Do not forget, the major aim of field trials is to **mimic what would happen in commercial agriculture**. Therefore field trials should usually be planted at the same time that the crop is normally planted. Planting depth, plant density, nutrient management, weed control, disease control, harvest time and method and post-harvest treatment should all match commercial production as far as this can be achieved within the restraints of small plot management.

Choice of land

In order to choose a good area of land for field plots it is necessary to identify the factors that magnify soil differences and to reduce, if possible, soil heterogeneity.

Fertility gradients are generally more common in sloping land. Soil nutrients are soluble in water and tend to settle in the lower land areas. Therefore these lower soils tend to be more fertile than the higher areas. An ideal experimental site will be on flat land but this is not always possible. For example, how many fields have you seen that are as flat as a football pitch?

If the land has previously been used for plot experiments, then this can lead to increased soil heterogeneity. Therefore areas of land that have previously been planted to different crops, different and varied fertility regimes, or subjected to varying cultural practices, should be avoided, if possible. In cases where this has occurred, then the area should be planted with a uniform crop, with uniform management and fertilization for at least two years before it is reused for plot experiments. A second source of soil heterogeneity is related to unplanted alleys or roadways from previous experiments. If possible unplanted alleys from previous research should be marked and avoided.

Grading (ground levelling) usually removes soil from elevated areas and redistributes it to the lower areas. This operation, which is designed to reduce slopes, results in uneven depths of top-soil and often exposes unfertile subsoil. These differences can prevail for many years and should be avoided unless soil heterogeneity trials determine that the grading effect is minimal.

Large trees and other structures can cause shade, which will affect plant performance, and also their roots spread further than their canopies and so will influence plant growth. Areas near buildings may be affected by soil movement and heterogeneity caused by the building operation. Plots adjacent to trees or wooded areas can also carry a greater risk of damage by birds or mammals.

The evaluation of soil heterogeneity requires growing **uniformity trials**. These involve growing a single cultivar (or a number of cultivars) in plots with very high levels of replication. Uniformity trials result in determining soil fertility gradients and identify particularly productive or non-productive areas in fields. Uniformity trials can be used to produce contour maps of productivity. Statistical routines such as serial correlation studies or least mean squares between rows, column and diagonals can be applied to determine significance of soil heterogeneity.

Although uniformity trials have their place in field experimentation they usually have little to offer a plant breeder. Uniformity trials indicate the response of specific genotypes to a given area in a given season. When these trials are repeated with different genotypes or in different years then a different result is often obtained (not surprisingly). In plant breeding evaluation trials, the number and diversity of genotypes under test are usually far greater than what can be considered in uniformity trials. Also it should be noted that often there is little choice of what land can or cannot be used for plot trials.

A plant breeding programme usually uses a number of different locations. One main location may be identified where the majority of material is evaluated in the early and intermediate selection stages or where seed is increased. A number of different locations will be used (dispersed throughout the region where the new cultivars will be targeted) where advanced lines are tested for adaptability. Where many locations are used it is common to use farmers' fields for test plot evaluation. Some of the distinct differences between a farmer's field and the conditions, which would prevail at, say, an experimental research station, would include:

- Lack of experimental equipment or lack of small plot machinery. This can often be easily overcome by taking planting, spraying and harvest machinery from the research farm.
- Lack of experimental facilities such as precise irrigation control and pest or disease control, weather stations etc.
- Lack of post-harvest storage or assessment facilities. Therefore harvested produce needs to be transported to a central testing laboratory for post-harvest quality assessment.
- Large variation between farms and fields within farms. This is often not a major problem as the majority of trials on these farms are to select for such adaptability over a range of environments.
- The farm sites are usually further away from the base research laboratory and sometimes long trips are necessary to visit the plots. Therefore visits are usually limited and it can be difficult to identify potential problems as they arise and hence avoid their worst effects.

Despite all the potential difficulties with off-station or farm trials it is possible to achieve very good results. Best results are usually obtained when the 'better' farmers are chosen for the tests and when these farmers are specifically interested in the results from the trials. Finally, when trials are to be carried out on farmers' land it is always advisable to keep the experiments simple and to have relatively large plot units, and this make them 'more' robust.

Plot size and replication

It is always assumed that larger plots are more efficient and more representative than small plots in yield and other assessment trials. Similarly there is no doubt that greater replication levels are always more desirable than fewer replicates. The difficulty of organizing efficient field trials is often related to some compromise in plot size and replication which will allow large numbers of test lines to be evaluated at low cost and in as small an area of land as may be available.

Land availability may not be the limiting factor in determining plot size or replication level. It would be pointless to organize more field plots than could be effectively managed by the staff available. Similarly, data needs to be collected from effective field trials and if too many unit plots are grown than it may not be possible to effectively evaluate either plants or the produce from the trials. Finally, some crop species produce products that are bulky or perishable. It may be necessary to store the produce (or at least a sample of produce) from each plot and the storage space available would then be a major determining factor.

In the early selection stages the amount of planting material available is often limited and this puts practical constraints on the field trialling that is possible. For example, if only 2 g of seed are available for evaluations, and commercial seeding rates are 4 kg per acre then only small plots with limited replication will be possible.

Increasing replication will always be more efficient than increasing plot size. Therefore if 200 plants were to be grown for evaluation purposes, then the most statistically efficient design would involve 200 replicates of randomized single plants. From a practical standpoint this may not, however, be the most effective or practical or provide the most representative outcome. For example, there may not be the necessary machinery available that would allow for mechanized planting of completely randomized single plants. Therefore the dimensions of machinery available can be a determining factor when setting plot dimensions. If the only plot seeder available plants six rows, then all plots are likely to be a factor of six rows wide. Similarly if a small combine harvester is available that has a cut of 1.5 metres then plots are likely to match this harvesting capability. In addition, single plant evaluation may take greater land areas that would not be available. Finally, single plants, if completely randomized, need to be spaced distinctly apart to differentiate one from another. The phenotypic performance of some crop species is markedly different when grown at wide spacing (wider than would be normal for commercial production) than if grown at narrow spacing.

A breeding plot can consist of a single plant, a single row or multiple rows. The plot dimensions are often determined by the availability of planting material.

Different plots in field trials invariably contain different genotypes. The performance of these genotypes can, in some cases, be affected by competition from the adjacent plots. For example if a short genotype is grown next to a tall vigorous genotype then the performance of the short type may be reduced compared to a single stand of the short stature plants. To a large extent these effects can be reduced by good experimental design and replication where the probability that adverse or advantageous competition occurring in all replicates is reduced with increasing replication.

Some researchers suggest growing larger plots and harvesting or evaluating only the centre rows (i.e. that portion that is completely surrounded by plants of like type). It should be noted, however, that this would require greater amounts of planting material and larger land areas. It should also be noted that genotypes can suffer as much (or greater) competition by being grown by itself and **ripple effects** can occur. To examine ripple effects consider a five row plot (rows A, B, C, D and E) where row A is grown adjacent to a different tall and very competitive genotype. In this case then the A row may contain small stunted plants due to the competition from the tall genotype and hence will result in lower yield. Row B, however, is likely to be affected by competition because although grown next to a like genotype, the like genotype (A row) is stunted and low yielding. Therefore row B will be taller and more productive due to the lack of competition from row A. In a similar manner, row C will have to compete with the larger more competitive B row plants and have reduced yield. The competition effects will be reduced, however, with increased distance from the tall different genotype and hence the name ripple effect.

It should also be remembered that by harvesting only a portion of the total plot the error variance will be increased as the error variance of the mean (average of all plants in the plot) is the error variance of a single plant divided by the number of plants.

It should be remembered that the value of field plot trials is to make comparisons and not to estimate definitive yield performance. Therefore field trials are used to compare the **relative performance** of different test lines in comparison to control entries. In this case increased or decreased yield as a result of competition will only become a factor if there is interaction between edge effects and genotypes.

Guard rows and discard rows

It is common practice to surround trials (and sometimes even individual plots with **guard** or **discard rows**. These are areas planted to a specific cultivar or genotype, which is not part of the evaluation test. Guard rows are used for several reasons, including:

- If any mechanical damage occurs (e.g. a tractor spray unit accidentally runs over a plot), it is likely to happen to the edge plots. If these are to be discarded, then this damage is less likely to affect the performance of any of the test or control entries
- Phenotypic performance can be greatly increased by avoiding differential edge effects. Therefore plots that are grown on the edge of a trial will not have any competition on one side, while all other test entries will be affected by competition from adjacent plots
- In multi-factor field trials, guard rows can be used to separate different treatment factors that may be difficult to apply to specific areas without having some effect on the immediately adjacent plot

Guard rows are usually the same species that are under evaluation, but this is not always a necessity.

Machinery

Over the past decades there has been an increase in the availability of small scale machinery suitable for field plot trials. Most of the machines are designed as miniature versions of what is used in larger scale agriculture. Tasks, which can now be mechanically orientated, include:

- Planting
- Weed, disease and pest control
- Harvesting

It is always desirable to plant field trials mechanically as this is likely to result in more uniform plots than can be achieved by hand planting. This is almost always true for small, and relatively small, seeded crops (e.g. barley, wheat and rapeseed). When the planting material is larger (e.g. potato tubers) hand planting can produce as good, or better, results compared to mechanical planting. The need for automatic planting will be dependant therefore on the size of seed to be planted, the density of seed sown and the time that can be saved by automatic planting.

The most common small seed plot planters are **cone planters** (Figure 9.6). This type of seeder can be used

Figure 9.6   Planting yield assessment trials using a single cone planter.

very successfully to plant either very small plots or much larger plots. Pouring a measured or counted quantity of seed over a cone such that the seed is evenly distributed around the base of the cone operates the seeder. During planting the cone revolves and seed passes through a hole and is subsequently dropped via disc or tube coulters into the soil at the required depth. It is often possible to plant several rows from the same unit seed lot. In this case, after the seeds drops from the cone they are evenly distributed to a number of tubes, which will each plant a single row. Cone planters are usually designed so that a range of plot lengths is possible. This is achieved by gearing the rate that the cone revolves. After one complete revolution then all the seed from one lot will have passed down the open hole. Planting can be done with continuous movement with each plot being dropped onto the revolving cone at a designated **trip point.**

Cone planters are available where the seed for each plot/row are loaded into a cassette or magazine. This is then mounted above a seeder unit with several revolving cones. With this system it is possible to plant several rows simultaneously with each row being a different genotype. Cone seeders are particularly useful as they can be used with small seed lots and all seed loaded is planted to completion. Therefore there is no need to maintain a seed reservoir, which needs to be emptied between different plots/genotypes. Cone planters are also self cleaning.

With small seed, cone planters can result in an even distribution of planted seed, but it is sometimes desirable to have a more precise placement of seed. If this is necessary then **precision planters** can be used. With these machines it is possible to obtain spaced plants at relatively even density. Precision planters are in general of three types:

- Belt planters, where a reservoir of seed is maintained over a revolving belt. The belt has holes cut which are of precise size and shape so that only a single seed will fit through the hole. The density of planting is achieved by the number of holes in the belt and the rate of which the belt revolves (e.g. Stanhay seeders).
- Vacuum planters, where suction is applied to a revolving plate that has holes drilled to allow only a single seed to be sucked to the plate at each hole position. As the plant revolves the vacuum is turned off at a

specific place in the plate's rotation. At this point the single seed is dropped into the soil.
- Cup planters, where a series of rotating cups are dipped into a reservoir of seed. The size of each cup is such that only a single seed is scooped up as the cups revolve through the seed. At a specific point on the rotation the cups are tipped and the single seed is dropped into the soil.

The major limitation of precision planters when used by plant breeders is that they usually require a volume of seed in the reservoir in order to operate effectively. Therefore they have only limited use when small amounts of seed are available.

In some crops, transplanting is common even on commercial scale (e.g. fresh tomatoes). Small scale transplanters are available which will allow automatic transplanting of field plots. Seedlings are grown in '*seedling flats*'. At transplanting time the seedlings are removed from the flats by hand and placed into the transplanter. Systems have also been developed where the seedling flat fits onto the transplanting machine and the whole operation is automated. In this latter case it is usually possible only to transplant large plots.

The areas between different plots in field trials are usually left unplanted. There is very little competition in these areas and weeds can be a big problem. Weed control in field plot trials can be carried out mechanically or chemically. Mechanical weed control can be by hand hoeing (a task often enjoyed by many summer student helpers). Automatic mechanical devices such as rota-tillers and harrow cultivators can achieve inter-row and inter-plot weeding. Often it requires a combination of chemical herbicide application, rota-tilling, harrowing and hand hoeing to ensure that plots remain weed free.

Evaluation of disease and pest resistance is an important factor of field testing. Test lines and controls will be grown in regions or areas where specific diseases or pests are common. Disease is often encouraged by including particularly susceptible genotypes as **spreaders** and by artificial inoculation of these spreader lines.

In other field studies it is not desirable to have disease or pest epidemics and this need to be controlled. Control is usually by chemical application although some biological control of insects may be available. It should be remembered that many diseases are spread (and most are not helped) by having poor weed control.

Figure 9.7   Small-plot combine harvesting rapeseed trials.

A variety of harvesting machinery is available including small plot combined harvesters (e.g. Hege or Wintersteiger) that will cut, thrash and partially clean seed samples (Figure 9.7) and harvesters that will dig root crops such as potatoes. Often very small plots (or individual plants) require to be harvested separately. In some cases this can only be achieved by hand harvest (e.g. pulling single plants or hand digging individual produce). In the case of grain crops the small plots can be hand harvested but the seed removed from the selected plants by small-scale mechanical thrashing.

USE OF COMPUTERS IN PLANT BREEDING

Many routines in a plant breeding programme follow a cyclic annual operation. Therefore the same tasks (or similar operations) are carried out on a seasonal basis. In general terms a simple breeding scheme may involve:

- Deciding which breeding lines are to be tested. Which lines are to be tested and in which environments? Which characters are to be evaluated from each trial? What control or check genotypes will be included in each trial for comparisons?

- Designing experiments or evaluation trials. What types of experimental design will be used for each trial (unreplicated designs, randomized complete block designs, lattice designs and split-plot designs). Having decided on an appropriate design then field plot plans need to be produced, planting material organized and arranged in order for planting.

- A number of clerical tasks will be required such as producing plot labels or harvest labels (Figure 9.8); genotypes lists and perhaps score sheets.

- After field (or other) trials have been planted then data will be collected throughout the growing season, at harvest and post-harvest.

- As data are collected, each variate assessed needs to be analyzed. More detailed analysis of over-site trials and analysis to examine relationships between variates will be carried out.

- Statistical analysis is only one step in data interpretation. Further data examination techniques of scatter diagrams; histograms or bar charts can be used to obtain a better understanding of data.

- Selection will be applied based on information collected. Breeding lines will be divided into various categories (e.g. definitely select and advance to next stage, not quite sure so repeat in smaller trials, discard from the breeding scheme).

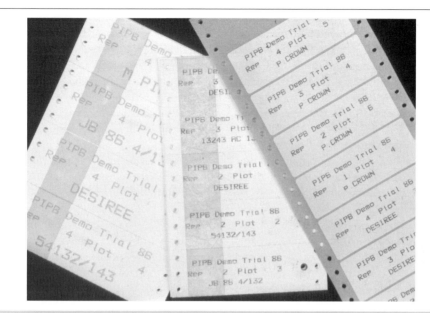

Figure 9.8   Assortment of planting and harvest labels used to organize breeding material.

- Once the 'best' genotypes have been identified the operation will begin at stage one again the following year.

This cyclic operation will continue over a number of years. Starting in the early stages where perhaps many thousands of lines will be tested at limited sites and with few characters recorded. Moving to an intermediate and advanced stage until after several selection rounds only one or two potential new cultivars have survived for varietal introduction.

Computers can be of great benefit to plant breeding in carrying out all of the tasks above, and perhaps even in others not mentioned. Often it is necessary to use different software packages for different aspects of the programme although there are a few packages that have been specifically designed for managing plant breeding programmes or for field experimentation studies.

Computer software packages that are available are all roughly of the same form. There is a central data storage (database) that can be accessed by a number of routines. Each routine will perform specific operations. Various routines may add information to the database while others will take information from the database and carry out a specific task.

The following section will examine a number of the options or routines that are available and explain how they may be used to increase the efficiency of a plant breeding programme.

Data storage and retrieval

All plant breeding schemes will generate vast bodies of data. If these data are to be used effectively for selection of the most desirable genotypes then reliable storage and retrieval of information is essential. Computers offer the option of storing data in such a manner that data sets can be tabulated for inspection in a number of different ways using **database management systems**.

In simple terms there are two types of databases used in plant breeding called **breeding line databases** and **germplasm databases**. There are a number of differences between the database structure depending on the two types.

Plant breeding databases will store information on assessment trials. Therefore, a large proportion of the entries in these databases will be discarded after each selection stage. Early generations will have thousands of **records** (where one record is associated with information from a single genotype) with only a few data scores on each. Conversely, a genotype, which survives to the advanced stages, will have been assessed over several years (and in many of these years, assessment will have been carried out at a number of different locations). Therefore the amount of data storage space needed for each record will depend on the stage that has been reached in the breeding scheme.

Germplasm databases have several differences compared to plant breeding databases. Germplasm databases hold information on a wide range of different genotypes. However, unlike a breeding database, new accessions (or records) are added but very rarely are records deleted. Information stored in a germplasm database will have been collected over years and sites but not all accessions will have been assessed in a common environment. It is therefore necessary to **rate** accessions (e.g. on a 1 to 9 scale or A, B, C, etc.) so that comparisons can be made. It is vary rare that actual yield data (e.g. t/ha) are stored on a germplasm database it is more likely that a particular accession will be rated as a particular 'score' for yield.

Irrespective of the type of database, each record (test entry or accession) will be assessed for a number of characters or traits of interest. Variates can be of two forms, **numerical** (e.g. disease rating of 2, yield of 25.32 t/ha) or **character** (e.g. alpha/numeric character string like 'Yellow flowers'). In addition the different database types will hold other information not related to simple assessment, for example, an alpha/numeric string to identify the particular genotype (e.g. PI.23451 or 89.BW.11.2.34) and parentage of the line. Germplasm databases may store information not usually stored on a database for example, species name, ploidy level, source of origin of seed, age of seed, amount of seed available etc.

When a particular genotype entry is introduced into a plant breeding programme it is usually identified by an alpha/numeric code. For example, cultivars will have specific names 'Jack's Wonder'. Genotypes, which have derived from other germplasm collections, will have an accession number. For example USA plant introduction lines all have PI numbers (e.g. PI.12342).

Different genetic lines, which have derived from a breeding scheme, will generally have similar identifying codes. Genotype codes can be assigned in numerical order (e.g. line 1, line 2, etc.). It is more useful to assign an identifying code, which provides some information regarding the background of a specific genotype. For example in a specific rapeseed/canola breeding group

all crosses made are assigned a code identifier which includes the year of crossing, a two letter code of the purpose behind the cross and a numerical number. A cross identified by 93.WI.123 would indicate the 123rd cross made in 1993, with the purpose of developing a winter industrial (WI) type.

Specific individual genotypic selections from the cross would have different trivial numbers (e.g. 93.WI.123.23, 93.WI.123.69, etc.). If some form of pedigree selection scheme is used then additional trivial numbers can be added to indicate the number of within population selections made.

In setting up a suitable database the user must decide on a suitable **database structure**. This will determine the number of entries, which are to be tested in each trial, the number of locations where evaluations will be carried out at each stage and the number (and type) of data that is to be stored.

Irrespective of the type of database or form of data storage the primary aim is the same, to make information available for inspection in a clear and concise form.

Field plan design

Field trials and experiments are of major importance in a successful plant breeding programme. The ability to use computers for randomization has been realized for many years and most programmes use some form of computer generation of field trials. These packages use entered information such as type of design, experiment title, number of entries, number of replicates etc. and produces a randomization along with a map representation of how the plots will appear in the field.

Once a computer has generated a field design it is possible to store all the trial details, number of entries, entry codes, actual randomization, on a database system. This information can be retrieved later for analysis of data or producing plot labels.

Clerical operations

Despite advances made in database management, the ability to carry out complex selection strategies or analysis using computers the simplest and most useful task a computer can do for a plant breeding programme is to perform as many of the routine clerical operations as possible. Several years ago, all plant breeding schemes produced all field maps, genotype lists, seed packet labels and score books by hand. Even in cases where highly methodical and dedicated staff are used there are inevitable transcription error, which occur when hand writing and more importantly it is time consuming.

Computer systems can easily be used for:

- Printing field plot plans (mentioned above)
- Printing a variety of labels that can be used to organize experiments, store seed or planting material or harvest experiments (Figure 9.8). Organization of large randomized experiments can be achieved with ease if computer labels are printed with the genotype identifier along with a field position (e.g. either a plot number or a two-dimensional array number)
- Printing score books, which show simply a number of boxes in which data, can be handwritten for a number of traits
- Genotype lists of lines, which have attributes in common (e.g. all bread quality wheat, all industrial rapeseed etc.)
- Data summary tabulations of information stored in the database
- Keeping track of exactly where each test entry is (e.g. what stage of the breeding scheme, what sites each entry is evaluated, exactly where in the field can each test line be found)

Data collection

A breeding scheme is only as effective as the information collected on how the different genetic lines perform. The breeder must collect data on performance of different traits from appropriate assessment trials. Data collection and data management are areas, which have received very little attention in a plant breeding context. However the significance of the information-gathering stage is of great importance.

Data management is of three types:

- Collection of information into a form suitable for computer entry
- Validation of data to ensure that errors can be corrected before data analysis
- Sorting data into a form suitable for entry either into analytical software or for storage on a database

Data collected from experiments can be hand recorded onto score sheets and then entered (i.e. key

to disc) at a later stage. This form of data collection and entry may appear inefficient compared to more direct systems (below). However, there are one or two advantages of hand recording and later entry. There is always a hard copy of the information collected that can be referred to at a later date. Hand recording of visually assessed data can often be achieved quickly compared to other means (although the data still needs to be typed later). Therefore a combination of experienced assessor and experienced typist/data recorder may be as quick and efficient as directly logging data.

Information can be **logged** directly into a computer system. Data logging can be of two forms:

- Information in entered into a hand held data logger
- Information is transferred direct from an analogue/ digital machine (e.g. electronic balance or moisture meter).

In each case, the data are usually later transferred to the main computer system. Data validation (e.g. are the numbers reasonable or within a certain range) can often be achieved during, or as a part of, the transfer operation. Alternatively, data may be validated or checked as entered into a hand held unit.

Automatic transfer of data from analytical machinery is always an advantage as it reduces time and effort to achieve results. More important, however, is that this form of data collection usually avoids any additional transcription errors (e.g. writing down or hand typing the wrong number).

Hand held data loggers are rapidly becoming smaller and more sophisticated. However, it can often take longer to enter data (particularly alpha/numeric character information) into a handheld unit than to simply write the information. In addition, some expertise is required in the use of handheld units, particularly accurate typing skills. Finally, if data are collected in a handheld data logger it is always best to have some form of hard copy printout of the data as recorded. This would indeed apply to any data recording.

Data analysis

One primary consideration for the analysis of assessment trials is the ease and speed of processing. Often the most important traits (e.g. yield and quality) are not recorded until late in the season. A rapid throughput of analysis can therefore be critical to allow selection decisions to be made and new trails organized before planting. This can best be achieved if all the genotype identification codes, experimental design details and randomization information have previously been stored in a database. In order to carry out an analysis of variance for a single variate assessed at one location and produce an easily understandable but comprehensive output it should only be necessary to enter parameters to identify which trial is to be analyzed and the variate name.

As data are collected throughout the growing season analysis of individual traits can be carried out soon after data collection. Inspection of de-randomized data and genotype averages can often serve as a good check that there are no major errors in the data. It is important that each variate be analyzed to determine the variability within the genotypes for particular characters. Most database systems will automatically store means and statistics as analyses are performed.

The mode of data entry will, to a large extent, be determined by the method used to collect data (i.e. automatic logging, data logging or pencil and paper). Irrespective of how the data are collected, eventually the data to be analyzed will be available for entry into an analysis and storage scheme. The order that numbers are entered can differ from one of no pattern (not a good idea), field plot order (either going across the trial or up the trial) or in standard order (e.g. genotype 1 replicate 1; genotype 1 replicate 2; genotype 1 replicate 3; genotype 2 replicate 1; etc.). It is important that data be entered in the order expected by the software package.

Other features which will facilitate a rapid and efficient turnover of analyzing individual traits and storing information will include:

- The ability to estimate missing values
- To analyze only a subset of the total number of replicates (e.g. in a four replicate trial data for some traits are collected only from replicates three and four)
- To be able to transform (e.g. ARCSIN transformation for percentage data) or convert (e.g. convert dates to days after sowing or to convert plot yield into t/ha) data before analysis
- To derive variates from single, or multiple, data sets before analysis (e.g. if grain yield is recorded along with straw weight, total above ground biomass can be derived by adding the two recorded characters)

If multiple environments are used (say at the advanced trial stage) then over-sites analysis (simple analysis of variance or joint regression analysis) can be performed using stored means from individual site analysis. If an assessment trial is grown at two (or more) locations, and yield per plot is recorded from multiple replicates at each site, the following procedure can be used to obtain an analysis of variance of yield over sites:

- Analyze data from each location separately and store the genotype means on a database, along with the error variance from the analysis
- When each location has been analysed separately then an analysis of variance with source terms: genotypes, locations, genotypes × locations and an error term can be produced quickly and easily. The error term is obtained by simply pooling the error terms from each of the individual analyses

To interpret data from assessment trials and provide indications of possible selection strategies then joint regression analysis, over-site analysis, simple and multiple regressions and correlation analysis can all offer an insight into the variability of characters and also the relationship between traits. In addition, visual inspection of histograms and scatter diagrams can help in decision making. Multi-variate transformations (canonical analysis, principal components analysis etc.) have been suggested as possible aids to plant breeders by reducing the dimensions of selection problems. If these transformations are readily and easily applied to breeding data sets perhaps plant breeders will more readily use them.

Alongside complex analysis it should be possible to carry out simple calculations. Simple calculations would include addition, subtraction, multiplication and division. Other calculations, which may be helpful, would include expressing data as a percentage of either the trial mean or the average performance of one or more control line.

Selection

If many hundreds of lines are to be considered for selection, then computer simulation (by selecting a subset and comparing that subset to those lines rejected) can be a big help in either setting culling levels for different characters, or in setting weights in an index scheme.

The speed that different selection strategies can be compared using computers offers the potential of investigating a number of different selection options within a narrow time schedule between final assessment of genotypes and preparations for planting the following stage trials.

Data transfer

The amount of data collected on individual genotypes in a plant breeding programme is directly proportional to the stage of selection. By the most advanced stages, data from surviving lines will have been collected over several years and locations. If plant breeding database systems are to be of a useable size and if all information available is to be stored together in a common database then each season either:

- Records for discarded breeding lines must be deleted from the database and new data storage allocated to those selected lines, or
- Selected lines must be transferred, along with any data collected, to a new database where additional storage is available

Either option can be used and both are equally efficient. If the first option is chosen the old database can first be copied before the unwanted records are deleted. This allows access to data from discarded lines, which can often be used in the future, for example to gain an indication of particular defects of specific parents.

Statistician consultation

It is essential that agricultural experiments have clear objectives and that they are well organized and designed based on sound statistical reasoning. Consultation with qualified statisticians should be done before, and also throughout, the experimentation period. Plant breeding assessment trials are no exceptions. There has been some concern that statisticians will not be consulted if breeders are capable of easily generating a number of different experimental designs and performing complex analyzing of data from these experiments. Care must be taken to ensure that the appropriate design is chosen to answer the questions required. Most plant breeders' trials are, however, of a standard form where a number of test genotypes are compared in performance with a number of standard or control cultivars. Although the majority of plant breeders are more than capable of using the appropriate experimental design and making the

correct interpretation to standard analysis it should be noted that many analysis types (e.g. multi-variate analysis) are now readily available to non-qualified workers but that interpretation of these results often requires an experienced person. The point is, therefore, that statisticians should not be ignored and where possible they should be consulted and encouraged to contribute ideas in data interpretation.

Ease of use

One feature about computers and computer software that has not been discussed is the ease of operating the system. Many software packages are **user friendly**, which means that they can be used by relatively inexperienced staff. This does not however, imply that these database systems can be used without computer training. There will be at least minimal training required if a database scheme is to be integrated into a breeding programme.

Most user friendly computer packages give clear and precise instructions in the form of prompted messages to which the user replies with one or more operations or data entries. In many cases these prompted instructions can partially eliminate the need for 'user manuals'. It is, however, general experience that a combination of prompted commands along with a fully documented and a concise users' manual will normally be required.

RELEASE OF NEW CULTIVARS

The ultimate goal of any plant breeding programme is to develop superior genotypes and to release these into agriculture as new cultivars. The final stage of a breeding scheme therefore involves the process of release, perhaps protection, and distribution of planting material to the seed industry and/or the farming community.

The first part of this process is when the breeder decides that a particular genotype has merit as a new cultivar. This decision will have been made having observed the performance of the potential new cultivar as it passes through all the stages of the breeding programme. This would entail a number of years and in, the more advanced stages, and a range of different locations. It cannot be stressed too strongly that if there is any doubt regarding the worth of a potential cultivar then these doubts must be addressed before deciding to release a new variety. The general agricultural community does not generally take kindly to being sold seed of a cultivar which proves to be of little, or no, use. In the seed market a good reputation is difficult to obtain but easy to lose!

In most cases the decision to 'release' or 'launch' a cultivar is not the exclusive decision of the breeder. Where a breeder is working for a commercial company, then the final decision to take the first steps towards commercialization is unlikely to be made only by the breeders. Others within the company, board of directors, financial marketing staff etc. will all contribute to the decision concerning the potential commercial impact that the cultivar may have, and more importantly, the potential profits that can be expected to the company if release is successful. If the new cultivar has been developed in a University department or other public organization then the final decision on release may involve heads of department and deans of the college or experimental station. Irrespective of whether public or private investment has financed the development of the line then there is logic in the breeder having a major input on the final decision.

In this decision making process, the requirements (often statutory) that are made of a new cultivar must be borne very clearly in mind. If the cultivar fails to meet the stipulated criteria then it will not be possible to commercialize it and all the effort will have been wasted.

Information needed prior to cultivar release

Distinctness, uniformity and stability (DUS)

Before a breeding line can be considered for release it must be shown to be **distinct** from other cultivars that already exist. Distinctness can be for morphological characters (e.g. flower colour) or a quality trait (e.g. low linoleic acid content in the seed oil). It is sometimes possible to say that a new cultivar is distinct for a quantitative trait such as high yield but in this case the new cultivar must **always** express the high yield character if release is granted and in practical terms is not an easy way to proceed. More recently, breeders are using molecular techniques to distinguish new releases from already existing cultivars.

The new cultivar must also be **stable** and **uniform** (i.e. stable over several rounds of increase) so the genotype must always appear the same irrespective of where it is grown. Therefore if a new cultivar is released which

is described as having uniform white flowers then all individual plants grown must have white flowers.

Careful attention to the final stages of seed increase and meticulous care in producing breeders' seed can be of great benefit in ensuring the uniformity and stability of the new variety.

Value in release

Prior to releasing a cultivar, breeders must demonstrate (from data collected from evaluation trials) that there is indeed merit in releasing the new cultivar. This will involve presenting data from several years testing and from a number of locations but the exact requirements and procedures will vary from country to country.

In many countries, government authorities carry out independent testing of all new cultivars before release is allowed. These trials are carried out over two or three years and at many locations throughout the target region. The aim of these trials (National List Trials) is to ensure that new cultivars are suitably adapted for the region. If breeding lines show sufficient **Value for Cultivation and Use (VCU)** then they will be added to the National Variety List of the particular country. In the case of EU (European Union) countries, when any new cultivar is placed on the National Variety List of any EU country then it is automatically entered onto the EU Common Catalogue and can hence be increased and sold in other EU states.

In other countries, such as the United States, there are no regulatory National List Trials in which each new cultivar is evaluated. However, each US state has an appointed body of people who will review performance data for all new cultivars and determine whether they merit release within the particular state. Breeders can submit data for release in more than a single state simultaneously.

Cultivar names

Any cultivar, which is to be sold commercially, must be given a unique name (or identifying code) prior to variety release. Within any given crop species there should only be one with that particular name. So a wheat cultivar and a potato cultivar can both have the same name, say, 'Sunrise', but two barley cultivars cannot have a common name, say, 'Maltster'. Unless there is some unfortunate problem such as a cultivar has mistakenly been allowed a duplicate name, it is difficult to change the name after release.

Hybrid cultivars are often given a number code rather than a recognizable name. In such cases the number code has a prefix which identifies the company responsible for its development.

In choosing names it is useful to select ones that are easy to remember and convey, if possible, the right image (e.g. 'Star', 'Golden Supreme' or 'Bountiful'). Equally it is obviously wise to avoid names that are obviously inappropriate such as 'Usually Dies', 'No profit' etc.

Another point to bear in mind is, if a cultivar is to be marketed in a foreign country or in an area where a second language is common, then it is important to check that the cultivar name does not have an unfortunate meaning in the other language, that it does have a desired image in that language and that it be easily pronounced. For example, there is no 'w' in the Spanish language alphabet so it would be unwise to call the cultivar 'Wally's Wonder' if it is to be commercialized in a Spanish speaking country, such as Spain or Mexico.

Cultivar protection

Several USDA breeding groups offer the cultivars they develop, free from royalties, to the farming community (although the seed itself still costs money to cover production etc.) and therefore do not hold any rights on the new varieties (in other words the cultivar can be multiplied and sold by others). However, it is now very common to obtain some degree of proprietary protection of ownership on new cultivars that are released. All commercial companies require proprietary ownership of the cultivars they develop in order that they can control the supply of seed, who can grow the crop and to obtain either seed sales profits or royalties from seed sales.

When the cultivar produced is a hybrid, then it is possible to have 'automatic' proprietary ownership by simply maintaining the inbred parents that are used to generate the hybrid seed, and not allowing access to seed of the parents. In the end, if a cultivar is not a hybrid then the most common method of protecting cultivars that are propagated by seeds is to apply for **Plant Variety Protection** (PVP), within the United States or **Plant Variety Rights** (PVR) in some other countries. Also, in the United States, when a clonal crop, or other asexually propagated crop cultivar is released then protection can be obtained by **Plant Patent** (PP).

In each case PVP, PVR or PP allows the developers of a new cultivar to control how much, where and who has the right to grow and sell the proprietary cultivar.

It is as well to be aware of the information that is required to submit an application for either protection type, before starting out on the trialling and selection procedures!

Patents

In fact there is a much more general consideration of patents as they affect plant breeding which has recently become a major consideration in a number of ways. For example, it is possible to patent DNA sequences if they have a known function, it is possible to patent processes and methodologies, so for example tissue culture protocols or transformation methods. There are basic criteria that apply to patent applications but in essence you can apply for a patent if the 'invention' is not patented already, you can demonstrate a significant novel step and it is 'not obvious'. Clearly these have definitions attached to them but nevertheless this underlies the main principles. So increasingly, if researchers make any sort of breakthrough they will tend to consider patenting to protect their Intellectual Property (IP). In other words they can potentially make money from exploiting that technology either directly by exploiting it and preventing others from doing so or indirectly by licensing others to allow them to use the technology, perhaps from royalties. This affects quite a number of protocols related to plant breeding – particularly related to biotechnology and molecular biology.

The patenting of gene sequences also has rather obvious effects on the freedom to simply use particular regions of DNA.

So you cannot, as a breeder or breeding organization, simply carry out all tissue culture procedures including many of the transformation protocols if you intend to produce a commercial product. This restriction also applies to many genes and gene constructs. If you simply carry out research there are few restrictions if you can persuade the appropriate organization to give you the material, but methods are fairly freely used on the basis of research only. For commercial, in the very broadest sense, application you have to come to an agreement with the patent holder.

There has been, and still is, continuing debate on the ethics of being able to patent DNA – some argue that it is the basis of life so cannot be patented while others claim that the work effort and technology required to identify and extract it, means they must have protection so that they can recover their investment. Other controversies have been raised over who has rights when for example a gene is found in a species which has been collected as germplasm from another, perhaps developing, country?

Thus some years ago breeders worked on a principle that '*pollen was free*' but specific genotypes were basically protected (i.e. you could use any material as a parent but could not exploit it directly). Now this has been overtaken by, for example, protection for a number of processes the breeder might consider using; a genotype has to be changed substantially before being recognized as still not being the property of the originator; and paying of royalties to include a particular gene in a cultivar for exploitation. There are, of course, no 'black and white' answers or solutions but it is an increasingly complex area to operate within.

GENETICALLY MODIFIED CROP PLANTS

Clearly there are biological, agricultural and practical considerations associated with the use of Genetically Modified Organisms (GMOs) and these have been touched on earlier. However there are also other issues that this new and exciting technology raises. These issues include the concerns and worries that have been raised in relation to scientific issues, risk assessment, public concern and more general social aspects. It will not be possible to cover all these here, nor would be appropriate to try to encompass all that was necessary in this forum. It is perhaps worth quoting what might represent an attempt at providing a balanced view of the situation, it is from a 1998 UK House of Lords Select Committee stated in their 'Summary of Conclusions and Recommendations' that '*Biotechnology in general and genetic modification in particular offer great potential benefits to agriculture. . . . There are potential risks relating to environment, including the impact on the ecosystems of out-crossing, pest resistance*

and stress and multiple tolerances. We consider that environmental risks and benefits should be assessed at the same time.'

The advances in crop development by genetic engineering have occurred so rapidly that non-technical issues have primarily influence the initial introductions of these crops into agriculture. In most countries, genetically engineered plants come under the control of regulatory government organizations. Working with recombinant DNA techniques requires a degree of documentation and government approval to conduct such research. Permits need to be obtained before genetically engineered plants can be 'released' that is grown in the field, or greenhouses. Areas where field evaluations have occurred, often have needed to be monitored for a period of 2 to 3 years to make sure there are no detected 'or perceived environmental "risks" associated with the presence of the introduced gene'.

However, there have been a number of public perception issues over the use of genetically modified organisms, with a number of concerns being raised. Some have been based on ill-informed and irrational scares while others reflect the real uncertainty that surrounds any new technology – particularly one that centres on the essential feature of the life of any organisms – DNA. These concerns and the issues that surround them are important but cannot be given full justice here. But what must be stressed is that any breeder needs to take into account both the general prevailing scientific knowledge as well as the general perception of the issues involved. Both these aspects are changing rapidly at present – as might be expected with such a new technology. Breeders are particularly well placed to help clarify in their own minds, and hence inform others, of the issues involved. They need to keep up-to-date with what is being made possible by scientific discovery, what it might enable them to do, what risks it might present, what risk assessments need to be carried out, what benefits it might confer and what needs to be checked before they are happy to use particular genes or types of genes. They must clearly be governed by the scientific facts but they can also clearly identify risks and benefits from their particular intimate knowledge of the crop and its applications. But in the end, irrespective of the science, if consumers are not prepared to buy the product there is no point in the breeder proceeding!

As an example of one of the issues, most of our crop species have close relatives that exist as weeds in the same region as the crops are cultivated. One of the environmental concerns that have been expressed is that the genetically engineered traits will be transferred to these weedy species, which will mean that they will spread and be difficult to control. Indeed, investigation of this has shown that these transgenic traits will indeed be transferred to weedy species, even if not quickly, in the longer term it is probably inevitable. The question that arises is really how much of an advantage will the plants in wild populations gain from the presence of such a gene, and hence spread. Clearly when invading a farming environment in which the particular herbicide is used any genotype with resistance to that herbicide will have a distinct advantage and spread. However, when a herbicide with a different mode of action is used the advantage should be eliminated, unless the resistance is 'horizontal' and covers a number of different herbicides. In the situation where no herbicide is used, the persistence of the transferred gene will depend on whether it affects any other traits (positively or negatively in terms of natural selection) and particularly whether it carries a 'penalty' to its carrier. For instance, it may mean that the plant makes unnecessary use of metabolic energy producing something that is not needed. Similar scenarios will exist for other genes that are transferred into crop plants. The 'cost/benefit analysis', in the broadest sense of each is needed and any possible uncertainty investigated in detail. All technology carries benefits and costs – it is a question of the balance between the two, but must be heavily weighted in favour of safety to the consumer and the environment!

Development of the techniques, or their application, to produce transgenic crops has been costly and biotechnology based seed companies are anxious to recover their costs. It remains to be shown whether the advantages of transgenic lines will be sufficient to outweigh the cost incurred and whether investments are recouped. However, the 'weight' of investment must not cloud the scientific basis of any risk assessment.

Many of the early transgenic plants have resulted from genetic engineering of cultivars on which there is no plant variety protection (i.e. they have no legal owner). If transgenic crops are to have a large impact on agriculture then genetic engineering companies need to develop cultivars which are transformed for specific traits that cannot readily be produced by more traditional means and to work in collaboration with traditional breeding programmes/companies in order to keep development of the myriad of other characters increasing in performance, and in order to compete successfully in commercial contexts. But, at least in the immediate future, perhaps the greatest profits from genetically engineered crop species will be achieved by the legal profession who are likely to be called upon to sort out some of the difficulties that this new technology generates, particularly in the area of 'Intellectual Property' (IP)!

Plant transformation methods complement more traditional plant breeding work by increasing the diversity of genes and germplasm available for incorporation into crops and by (perhaps) shortening the time required for the development of new cultivars. Genetic engineering of plants also offers exciting opportunities for the agrochemical, food processing, speciality chemical and pharmaceutical industries in developing new products within crop species and offering new manufacturing processes. However, it is highly unlikely that these techniques will replace the traditional techniques that have been used in the past. Recombinant DNA techniques will, rather, add to the 'array of possibilities' to plant breeders in future cultivar development.

THINK QUESTIONS

(1) You are employed as barley breeder for a commercial company. At present the company does not have any greenhouse facility. You are trying to convince your peers that the breeding programme would benefit from having a greenhouse: List five uses you would have of a greenhouse in your breeding programme if one were available.

Great job (we hope!) you have convinced the board of directors to proceed with purchasing a new greenhouse. List five features you would like to request to have in the new greenhouse facility.

(2) A friend has just bred a new apple cultivar, which has good yield, fruit shape, storage quality colour and appearance. However the taste quality is somewhat lower than might be desired. Nonetheless, your friend has decided to release the cultivar and has asked you to suggest a possible name. What is

your choice of name and very briefly explain your choice.

(3) Briefly outline the major reason for including multiple location evaluation trials in a plant breeding programme. List four problems that may be encountered in organizing and carrying out 'off-station' trials.

(4) What is the main purpose of conducting field plot research in a plant breeding programme?

The primary field research station used by the University of Idaho is to be sold for housing development. To replace this research facility, the College of Agriculture has purchased a farm near Genesee called the Allroks Farm. Neither you, nor anyone else, has ever grown research plots on this previously commercial farm. List four factors that you would look for to identify an area to plant your breeding plots.

(5) Identify one aspect of a plant breeding-selection scheme where unreplicated designs (UR), randomized complete block designs (RCB) and split-plot (SP) designs could be the appropriate experimental design of choice.

(6) You have been employed as Senior Breeder at the *Hass Bean Breeding Co.* for several years and have had to carry out your breeding operation with very few funds. Due to a recent takeover of the company by the *Human Bean Seed Association*, you have been given a vast sum of money from the takeover. You have decided to use this money to upgrade the computer facilities used by your breeding group. To achieve this goal you have appointed a computer programmer to develop a computer plant breeding management. List the five main features that you would request from the computer package to perform and help increase the efficiency of breeding future new cultivars.

(7) List four operations carried out in a plant breeding programme that might best be conducted in a greenhouse. Briefly describe one problem of using a greenhouse in a plant breeding programme.

Index